EPISTEMOLOGY, METHODOLOGY, AND THE SOCIAL SCIENCES

BOSTON STUDIES IN THE PHILOSOPHY OF SCIENCE

EDITED BY ROBERT S. COHEN AND MARX W. WARTOFSKY

VOLUME 71

EPISTEMOLOGY, METHODOLOGY, AND THE SOCIAL SCIENCES

Edited by

ROBERT S. COHEN and MARX W. WARTOFSKY

Boston University

D. REIDEL PUBLISHING COMPANY

DORDRECHT : HOLLAND / BOSTON : U.S.A.

LONDON : ENGLAND

Library of Congress Cataloging in Publication Data

Main entry under title:

Epistemology, methodology, and the social sciences.

(Boston studies in the philosophy of science ; v. 71)
Includes index.
1. Science—Philosophy—Addresses, essays, lectures. 2. Social Sciences—Philosophy—Addresses, essays, lectures. 3. Knowledge, Theory of—Addresses, essays, lectures. I. Cohen, Robert Sonné. II. Wartofsky, Marx W. III. Series.
Q174.B67 vol. 71 [Q175.3] 501s [300'.1] 82–16159
ISBN 90–277–1454–1

Published by D. Reidel Publishing Company,
P.O. Box 17, 3300 AA Dordrecht, Holland.

Sold and distributed in the U.S.A. and Canada
by Kluwer Boston Inc.,
190 Old Derby Street, Hingham, MA 02043, U.S.A.

In all other countries, sold and distributed
by Kluwer Academic Publishers Group,
P.O. Box 322, 3300 AH Dordrecht, Holland.

D. Reidel Publishing Company is a member of the Kluwer Group.

Printed in The Netherlands

TABLE OF CONTENTS

EDITORIAL PREFACE

The last decades have seen major reformations in the philosophy and history of science. What has been called 'post-positivist' philosophy of science has introduced radically new concerns with historical, social, and valuative components of scientific thought in the natural sciences, and has raised up the demons of relativism, subjectivism and sociologism to haunt the once-calm precincts of objectivity and realism. Though these disturbances intruded upon what had seemed to be the logically well-ordered domain of the philosophy of the natural sciences, they were no news to the social sciences. There, the messy business of human action, volition, decision, the considerations of practical purposes and social values, the role of ideology and the problem of rationality, had long conspired to defeat logical-reconstructionist programs. The attempt to tame the social sciences to the harness of a strict hypothetico-deductive model of explanation failed. Within the social sciences, phenomenological, Marxist, hermeneuticist, action-theoretical approaches vied in attempting to capture the distinctiveness of human phenomena. In fact, the philosophy of the natural sciences, even in its 'hard' forms, has itself become infected with the increasing reflection upon the role of such social-scientific categories, in the attempt to understand the nature of the scientific enterprise. One may say that whereas the earlier decades were marked by an attempt to model the social sciences on the natural sciences, the decades from 1960–1980 were marked by the reverse trend: how were the typical concerns of the social sciences with human action, rationality, social and historical practices and structures, to be brought to bear upon the characterization of the natural sciences themselves? How were causal and statistical models of explanation to be tempered by considerations of freedom, choice, social values and interests, in understanding the growth of knowledge itself?

The papers in this volume of the *Boston Studies*, culled from those presented to the Boston Colloquium for the Philosophy of Science during 1973–1980, address the range of these questions in a variety of ways. Paul Diesing asks "How can we overcome the influence of ideology on historical fact?" He gives a careful analysis of a long-term case-study of the historical reconstruction of some major political decisions (e.g., the Cuban Missile Crisis). Leon Goldstein also addresses the question of the methodology of historical investigation, in a study of "the character of the intellectual procedures implicit in the production of the kinds of claim to knowledge which are the outcome of historical investigations." Carol Gould, after a critique of causal models of explanations of human action, proposes a novel view of human agency as causal, and focuses on the need for a model of social interaction, in terms of reciprocity. Marjorie Grene slays the dragon of empiricist psychological atomism, proposing an alternative 'structural pluralism' in an account of visual perception which preserves both objectivity and the normative character of human vision. Ian Jarvie tells us that sociology is alive and well, not *despite* its multi-paradigmatic methodological variety, but *because* of it, and he offers in evidence a lively and critical review of major contemporary sociological theories. Spiro Latsis counterposes to some standard views of the rationality principle in the social sciences (including Popper's) an alternative rationality for the understanding of human behavior as neither caused, nor random, but as 'plastically controlled' by situational analysis. Latsis proposes a way to save the ideal of deductive explanation in the social sciences by providing validity to rationalistic explanations of a richer, more flexible sort, which are cognizant of the parameters of human agency and choice. Werner Leinfellner examines the paradigms and theories of Marxian economics in great detail, in terms of contemporary microeconomic analysis, and restates the ethical content of Marx's theory of alienation in a rigorous way. Hillel Levine, in a fascinating excursus in the history of science, inquires into the Jewish responses to the Copernican revolution, examining the work of Jewish astronomers and cosmologists, and the reactions of the Jewish community to the new scientific age. Lawrence Slobodkin deals with the unique evolutionary strategies of the human species in developing a 'normative, introspective self-image'. He argues that human evolution radically departs from the functional restrictions of animal life by means of such strategies. Langdon Winner, in a critical study of the philosophy of technology, points to the need for a normative approach which considers both the practical and the emancipatory parameters of technology.

The essays, however different in theme and orientation, exhibit a common concern for rigorous philosophical approaches to the human sciences, a concern to preserve both the requirements for scientific rationality and the goals of freedom, meaning and social values, without which the social sciences remain inadequate in their understanding of human phenomena. We believe that these essays represent some of the most intelligent and innovative work in the ongoing renewal of the philosophy of all the sciences.

Boston University
Center for Philosophy and History of Science
November 1982

ROBERT S. COHEN
MARX W. WARTOFSKY

PAUL DIESING

IDEOLOGY AND OBJECTIVITY*

This paper is an interpretation of how ideology influenced one research project, and of how we tried to achieve objectivity despite or by means of our ideological differences. The project involved about a dozen people working together for three to six years in research on bargaining in international crises. We tended to think of our ideological differences as differences of location on a radical-liberal-conservative or left-right dimension, vaguely defined. However, in retrospect a more relevant difference was in our attitudes toward the U.S. and the Soviet governments, the main crisis antagonists in our research. Our attitudes ranged from loyalty to hostility toward the U.S. government, and we had both defenders and critics of the United States involvement in Vietnam. Unfortunately we had no loyal defenders of the Soviet Union and Soviet foreign policy; our attitudes ranged from near neutrality to strong hostility. We consulted published works that ranged over the whole spectrum of attitudes toward both the U.S. and S.U., a broader spectrum than we ourselves could represent.

Ideology was deeply involved both in our factual researches and in our theorizing. Our facts were historical facts, collected in about fifteen 100–200-page case studies of particular crises, and our theories were chiefly mathematical models of bargaining and decision-making.

To begin with facts: In writing our case studies we immediately encountered a situation familiar to historians, namely that writers with different ideologies give different accounts of the same event. That is, they not only give different facts, but they also give different interpretations of the same facts. How does this happen? Our experience as historiographers provides an answer. We found that we tried to work out a story that fit together and made sense. We collected facts from various trustworthy sources and tried to make interpretive connections between them. By 'connection' I mean (a) a temporal causal connection: Kennedy appointed Clay his representative in Berlin August 30, 1961 because of Adenauer's August 28 letter to Kennedy. I mean (b) intention or purpose: Kennedy appointed Clay in order to strengthen German morale, in part. I mean (c) the complementarity of an actor's various intentions, their unity in a general strategy or approach to his political task: Kennedy's middle-line approach, middle between soft and hard line,

R. S. Cohen and M. W. Wartofsky (eds.), Epistemology, Methodology and the Social Sciences, 1–17.

hawk and dove, involved balancing and harmonizing firmness and concessions, firmness on the Three Essentials for West Berlin, concessions on the reunification of Germany. Clay's appointment represented firmness, or politically a concession to his hard-line advisers like Adenauer and Acheson; Rusk's August 28 trip to New York to arrange negotiations with the Soviet Union represented softness, and politically a concession to his soft-line advisers like Stevenson. I mean (d) a coherence between an actor's interpretation of the situation, his interpretation of new information, his general bias (hard, soft, middle), and his strategy. Kennedy the middle-liner interpreted Khrushchev as pretty tough but also open to reasonable discussion at the proper time, and his communications to Khrushchev aimed at both presumed components.

Partly these connections appeared more or less directly in the evidence: a theme in Adenauer's letter to Kennedy, the danger of German neutralism, reappeared in Kennedy's first letter to Khrushchev, suggesting that Kennedy had taken Adenauer's warning seriously. Also Kennedy stated his reason for appointing Clay in discussions with advisers. Partly we intuited a consistency or coherence in an actor's interpretations, actions, and general outlook or bias. Partly we assumed or postulated facts that would account for or make connections between facts we knew, and then looked for evidence to support the postulate. For instance, references to the Ulbricht-lobby in the Presidium, Khrushchev's reference to pressures on him by colleagues, contrasts in speeches by Presidium members, and other facts, were connected by postulating a hard-line opposition to the Khrushchev faction in the Presidium. Partly connections were suggested by the sources of our data, the memoir writer or speech writer or historian, who himself interpreted the facts he presented.

We kept on collecting facts until we had a story that fit together and made sense to us. Where the story had gaps, or where facts conflicted or did not fit, we searched for further facts or re-checked the facts we had; where the story made sense we did not search further but merely accepted whatever additional evidence happened to come in. Of course we read all the secondary accounts available in our three working languages, but it was also possible to turn up more facts by tedious following of various leads, and the issue was which of these various leads to follow and how much energy to devote to additional search.

Ideology enters into this process at every step. It enters most prominently in the historiographer's attempts to reconstruct the outlook and purposes of an actor – Khrushchev, Ulbricht, Kennedy. In some accounts Khrushchev was a bluffing bully, trying to take something he knew did not belong to him

(West Berlin), but ready to retreat if his bullying was resisted. His ultimate purpose was to communize the world by military force. In other accounts he was trying to protect East Germany against West German aggression of various sorts; in still other accounts his main interest was in improving Soviet agricultural production, or in improving East-West trade. If he gave a speech emphasizing the nuclear might of the Soviet Union and warning the West to watch its step, his purpose in one account was to warn aggressive militarists like Franz-Josef Strauss that aggression against East Germany would not be tolerated; in another account it was to reassure his hard-line colleagues so they would tolerate his proposed cuts in the military budget and increase of the agriculture budget; in another account it was to frighten the U.S. into accepting some aggressive move he was plotting. For all these accounts, ideology provides the basic premise as to what Khrushchev was like: Communism, and therefore Khrushchev, is a totalitarian tyranny ruling by force; Communists are atheists and therefore evil and deceptive; Communist officials are politicians, so they engage in bureaucratic maneuvering like other politicians; Communists are workers with no economic reason to expand their territory or influence; the Soviet Union is a pseudo-socialist power but must make a show of defending Socialist countries against capitalist aggression to maintain its leadership status.

Given an ideological preconception of an actor's outlook and purposes, the historiographer notices and emphasizes facts that bear this out and questions or reinterprets or rejects facts that do not fit. For instance, if Khrushchev was an aggressive bluffer, his public and diplomatic statements proclaiming his interest in peace and *détente* must be interpreted as lies; his publicly expressed worry about atomic weapons in West Germany was a lie, because he *must* have known that the U.S. retained control and had put them there for defence only. Evidence to support their purely defensive nature comes from U.S. government statements, which are accepted as true. But from a different standpoint the truth and the lying are reversed; and from a third standpoint both governments might be telling the truth but underestimating the aggressive machinations of subordinates – the CIA, the military – or perhaps underestimating the dynamics of the security dilemma in which sincerely defensive moves by one side look aggressive to the other side and provoke retaliatory defensive moves.

I do not mean to say that the historiographer ignores or falsifies facts. Rather, he uses a double standard. Facts that fit into his developing story and help make sense of it are eagerly accepted, but facts that do not fit or that disconfirm his account are treated skeptically and perhaps not even noticed.

The skepticism extends to the writer who reports such dubious facts. For instance, if Nehru reported to Macmillan on September 13, 1961 that in his opinion Khrushchev, whom he had just visited, seemed to be mainly worried about West German military power and that his purpose was defensive and limited, this account will be eagerly emphasized by one writer – fact, evidence! – but treated skeptically or ignored as one man's opinion by a second. How reliable is Nehru if he can be so easily fooled by Khrushchev? Besides, we know he has a blind spot for Soviet aggressiveness; he failed to condemn the Soviet resumption of nuclear testing two weeks earlier.

The double standard applies also to the *search* for facts. A fact that will provide a connecting link or that will confirm an interpretation is eagerly searched for, but disconfirming facts are not searched out. After all, the writers who might provide those disconfirming facts are pretty unreliable anyway, and their accounts have to be checked and rechecked, so it would be a lot of work with very doubtful prospects for any results; and time is short, the case study must be finished.

But if disconfirming facts appear – and they do – they are accepted and put into the case study – as puzzles or as bases for possible reinterpretation of some actor's intentions. There can be several reinterpretations of an actor's intentions during the writing of a case study, as more and more facts come in. What is interesting is that all these reinterpretations stay within the bounds of the writer's ideology, in the instances I have examined. For example, on October 6, 1958, the Chinese announced they were halting the artillery barrage of Quemoy. Why? At one time I connected this message in part to a statement by Chiang, reported on October 5, that Quemoy was unsuited to be an offensive base. In July he announced that Quemoy was now a forward base for imminent mainland invasion; then came the Chinese artillery barrage which pinned his troops in their underground bunkers. In early October he took it all back, and October 6 the barrage stopped except for token reminders. In searching *The New York Times* and London *Times* for confirming evidence, I disconcertingly found evidence that suggested Chiang's remarks in the October 1 – October 5 period were directed to the U.S., not to China, and were in reply to Dulles' September 30 speech which Chiang disliked greatly. Chiang was telling the U.S. that Quemoy was a purely defensive base that the U.S. should protect as part of its commitment to defend Formosa. Interpretation disconfirmed. I then focused on statements in Dulles' September 30 speech which Chiang did not like, in which Dulles asserted among other things that the U.S. had no commitment to support any mainland China activities of Chiang and that Chiang's return to the mainland was most improbable. New interpretation: the Chinese studied Dulles' speech, decided that they had made their point about invasions and raids from Quemoy, and halted their artillery. In both interpretations a concession by the U.S. or Taiwan, a renunciation of mainland raids, plus U.S. military firmness in the Formosa Straits, induced China to end the artillery barrage. The basic interpretation has not been changed.

So far I have mainly reported on how ideology operated in the writing of our own case studies. I now postulate, on the basis of insufficient evidence, that similar mechanisms were at work for the political scientists and historians whose case studies we consulted. The evidence consists of the agreement between a writer's selection of facts, his interpretations, and the lessons he presents, and the consistent factual differences between writers of different ideology. I also postulate, on similar evidence, that the actual participants used the same mechanisms, both in their ongoing interpretation of events and messages, and in their memoirs. I also predict that the same processes are occurring in you, as you react to my statements with eager agreement or skepticism and dismay, or a mixture of the two.

Exceptions: I make exceptions for minor participants. Examples: Kennedy's press secretary Salinger (1966, p. 191) reporting that a certain Russian gave him a message for Kennedy from Khrushchev on September 23, 1961, or Adenauer's press secretary von Eckardt (1967) reporting on whom Adenauer consulted on a certain occasion. I have no reason to suppose ideology is relevant to selecting these facts.

How can one overcome the influence of ideology on historical fact? One possible strategy is for each individual scientist to resist and overcome his own biases, to treat all facts equally even though some fit his biases and others contradict them. For example, Possony asserts: "... purposes must be reviewed with a sense of realism. This means that facts are acknowledged even when they do not fit preconceived notions, and that factual interrelationships are interpreted objectively, in their entirety, and not on the basis of selected evidence" (Stefan Possony, 1964). Here is a call for objectivity; but shortly after these remarks were published, their author hallucinated a tri-continental people's war, planned and controlled in Peking. Mao had already given the signal to start this war when Johnson sent U.S. troops to Vietnam; this convinced Mao's opponents in Peking that the U.S. would fight to protect its vital interests, so they forced Mao to call off the war suddenly. The result was a split in the leadership in Peking, with Mao attempting to overcome his opponents by starting a Cultural Revolution, which failed (Possony, 1967). This is objectivity? This example suggests that objectivity cannot be reliably achieved by simply accepting an obligation to consider all the facts. Such an obligation runs directly counter to basic human cognitive processes (cf. Steinbruner, 1974, Ch. 4), even apart from ideology, so its chances of being effective are nil. Consequently objectivity cannot be simply an individual obligation but must be located in the organization of the scientific community to be effective.

Our attempted solution to the problem of objectivity was partly technical and partly organizational. First, we used a standard historian's technique of getting as dense a set of facts for a given time period as possible. We did this by consulting as many sources as we could find, and especially by searching for sources with different ideological backgrounds. Since different writers presented different facts, we could add them all together and get more potential facts than any one writer could even have known about. This also provided a cross-check on the various accounts. We also had the benefit of later works such as Macmillan's memoirs (1972) which added facts not known to earlier writers. Also the different accounts sensitized us to varying interpretations of the same facts, so we could detach ourselves from the first interpretation that occurred to us and could consider other possibilities. The varying interpretations also caused us to search for additional facts that would disconfirm one interpretation or point to a reconciliation.

In drawing connections between facts we used the principles of transmission time – that a message takes time to transmit – and reaction time – that a government takes time to react to a message, time ranging from two hours if the answer is predetermined to several months if the government is sharply divided and a change of policy is called for. For example, it is most unlikely that the Chinese message of 1 a.m., October 6, 1958 was a reaction to a Chiang statement printed in London October 5, a.m., or a statement by Ambassador Yeh in *The New York Times* October 4, a.m.; there is insufficient reaction time. Dulles' September 30 speech is just barely right for reaction time.

Our facts and principles enabled us to disconfirm some histories, including our own earlier versions, by showing that the connections they make between facts could not or probably could not have occurred, and that different connections involving facts they did not mention probably occurred. The error can then be attributed to ideological bias, that is the attempt to connect the limited facts one does know about into a meaningful story. I illustrate with three examples from September 1961. The problem here is to explain the cooling off of the Berlin crisis and especially Khrushchev's October 17 statement that negotiations looked promising and a settlement in 1961 was no longer absolutely necessary.

First, Stefan Dörnberg (*Kurze Geschichte der DDR*, pp. 460–461) asserts that the resumption of Soviet nuclear testing September 1 sobered the West and induced them to negotiate, thus allowing Khrushchev to call off the deadline. This account is disconfirmed by (1) statements in Schlesinger (1967) showing that Kennedy was interested at least by August 21 in beginning negotiations, and (2) statements that the testing angered Kennedy rather than sobered him, that he sent an ultimatum to stop on September 4 and an order to resume U.S. testing a few days later, and (3) that he continued his negotiation program with a September 13 proposal for talks at the UN. This suggests that the tests had little effect on Kennedy's interest in negotiations, though they might have increased the urgency of negotiations some. Second, Oran Young (*The Politics of Force*, pp. 204–207) asserts that Soviet nuclear testing did not succeed in "breaking the will of the West". Instead, he asserts that Western firmness, including the appointment of Clay August 30, an announcement September 9 of a U.S. troop buildup in West Germany, and displays of military force in late August, caused Khrushchev to lose hope for major gains and consequently to call off the crisis October 17. This account is inconclusively disconfirmed by a series of informal messages from Khrushchev in the period August 5 to September 29 in which he requested negotiations and hinted at possible concessions, including dropping the demand for formal recognition of East Germany; and by the estimates of Kekkonen, Nehru, and Sulzberger in that period that his aims were limited and defensive. If his aims were limited and he was open to compromise already in August, it is more plausible to assert that his hopes were raised, as he himself asserted in his October 17 speech, by Kennedy's conciliatory messages of late September and early October including especially his September 25 speech. Also reports of Khrushchev's reaction to Kennedy's July 25 speech suggest that his reaction to firm measures, had he noticed them in September, would not be to back down but to think more seriously about a separate peace treaty.

Third, Smith (*The Defense of Berlin*, pp. 302–317) makes the crucial event of this period Adenauer's August 28 letter to Kennedy, which Smith asserts caused Kennedy to wake up, harden his position, and appoint Clay to Berlin, thus discouraging the Soviet bluffers. Thereafter Smith recounts Clay's activities around Berlin from September 21 on. This account is disconfirmed by evidence that Kennedy was on the negotiations track both before and after August 28. Other evidence from Moscow shows that Khrushchev paid no attention to Clay's activities, and evidence from Washington shows that Kennedy paid no attention either. Smith thus focuses on an irrelevant sidetrack; his account is a fairy tale, a fabrication with a moral, as are the other two accounts mentioned. Incidentally all three fairy tales have the same moral: if the opponent is firm he will not break our will, but if we are firm he will back down. This is history?

With all these facts to build on, are our case studies objective? We did not think so. I was very suspicious of one or two of the studies that made little sense to me, including the Quemoy study, and I searched for facts that would clarify what really happened. Meanwhile others were checking my Berlin study, which *they* found incredible. They would tell me about facts which I thought irrelevant or puzzling or which came from very dubious sources, but I in turn was unable to persuade them that my interpretation of Soviet motivations and outlook were correct.

It turned out that even with a dense array of facts different interpretations were possible based on different assumptions about the motivations of a government. This then is the crucial point at which ideology enters into a case history. Additional facts can disconfirm crude fabrications, as in the three examples above, but they are then replaced by more refined fabrications. An example is Slusser's *The Berlin Crisis of 1961*, which combines skillful detective work as to which Politburo member was where on which day, with wild speculations about what happened at some meeting, based on uncritical hard-line assumptions about ultimate Soviet and NATO aims in Berlin. Still, Slusser's story is factually superior to the three fairy tales mentioned above, especially if one ignores his speculations.

What ideology does is to predispose one to a certain possible interpretation of a person's actions and block the appreciation of a different interpretation. For example, how is it possible to believe that the 1100 NATO troops in West Berlin could be a threat to the 320,000 Soviet troops in East Germany? If a Soviet spokesman were to assert this, his statement must be a crude excuse, a transparent attempt to deceive. The Soviet aim in 1959 must therefore have been aggressive; West Berlin was no threat to them. Conversely, could the West German government possibly believe that its subsidies to West Berlin, the fifty movie theaters along the sector boundary with reduced rates to East Berliners, the radio station with Free World news, the CDU subsidies to refugee assistance organizations which contacted East Berliners and helped them escape, were merely humanitarian generosity? Obviously not; the West German government was systematically trying to destabilize the East German government. Were the refugees "unstable persons" who were lured to West Berlin by traveling CDU agents and by the garish, repulsive, subsidized displays of consumer goods, automobile traffic, and lurid sex magazines along the Kurfürstendamm, or were they normal people voting with their feet for freedom? Questions of motivation like these are purely ideological; one answer will seem obvious and the other incredible (or some other combination) based on one's ideology. And these questions determine the interpretation of other facts such as the U.S. reinforcement of its troops September 9, 1961.

Why could we not use facts to disconfirm ideologically based assumptions about motivation? For instance, both Khrushchev and Kennedy explicitly said what their specific intentions and ultimate aims were, and their numerous statements in various media were internally consistent; doesn't that settle the matter? No; any such statement can be dismissed either as conscious deception, unconscious self-deception, or false consciousness. For instance when Khrushchev stated in his memoirs that the purpose of the Cuban missiles was to defend Cuba this could be a continuation of the 'cover story' devised in 1962 to hide the Soviet defeat. When Kennedy stated that his ultimate interest in Cuba was that it be free, this could be dismissed as self-deception about the operations of U.S. imperialism in Cuba up to 1959 and the arrangements the CIA was even then in 1961 making for the return of Batista. Other facts, such as the kind of missiles put into Cuba, are circumstantial, open to varying interpretations, and thus cannot conclusively disconfirm an interpretation of Soviet motives though they can make it more or less implausible. Conversely, facts about the past effects of U.S. political control in Cuba can be dismissed as irrelevant to understanding Kennedy's conscious intentions, though they can make the assumption that he knew what he was doing more or less implausible.

Consequently we found that with more and more facts, some disagreements about causal connections were resolved but disagreements about a government's ultimate aims and immediate intentions remained – in part. We did reach some agreements. We learned to accept each other's interpretations as possibilities or as one-sided exaggerations with some truth in them. We could agree that Kennedy may have *thought* he was trying to 'free' Cuba, and that Khrushchev could possibly have worried some about a second U.S. attack on Cuba. We could even combine two or three interpretations, asserting that all these factors were probably present in the case but we could not assess their relative importance. For instance when the S.U. put missiles into Cuba, how important was the defense of Cuba, how important was the improvement in the strategic military balance, how important was the challenge to U.S. resolve, how important was internal political maneuvering for pieces of the Soviet budget, and how important was the status struggle with the Chinese? We don't know. Were these partial agreements a step toward objectivity, or were they the product of small-group dynamics, of tacit bargaining, or of all three? I don't know.

I turn now to the effect of ideology on theory, specifically on the construction of mathematical models. Ideology operates first to give the modeler his sense of the problem, of what is to be clarified or explained or brought to people's attention. This sense of what is problematic determines his selection of the heuristic situation which will guide him in the construction of his model. Having selected a heuristic situation on an ideological basis, he lets his mathematical imagination play over it, selecting the essential relations to be included in his model. Here ideology operates a second time, to obscure or hide completely some things from his awareness and to bring other things to his attention. Consequently the situation on which his model focuses (the problematic to be studied) and also the postulates (the factors and relations brought to his awareness) are ideologically determined. And third, since ideology influences the construction and interpretation of a model, it also induces its inventor to become attached to it and to interpret reality with it. Officially its postulates are mere assumptions but actually its inventor believes they are true and important.

For example the early Ellsberg constructed a bargaining model in the 1950's, before he was awakened from his dogmatic slumbers by the U.S. war crimes in Indochina. This was a game model, of blackmail (Ellsberg, 1968). There are two players, the blackmailer and the defender. Play begins when the blackmailer demands an object having S_1 utility to the defender, under penalty of a war costing P_1 to the defender and P_2 to the blackmailer. The blackmailer may or may not be bluffing. The defender's task is to estimate the probability of bluffing, multiply this probability by P_1, etc., and thus to calculate whether or not he should resist. The model is Chicken. The heuristics underlying this model are obvious: the bluffing blackmailer is Khrushchev or Stalin before him, the prudent defender is the U.S., and the object of the grab is West Berlin, or North Iran, or diplomatic recognition of a non-state, East Germany. (See also Schelling's heuristic examples, in which the Chinese and the Russians are blackmailers: 1960, pp. 158–160.) Another example is the game of Chicken itself in Schelling's treatment of it (Schelling, 1966, pp. 116–125). His first heuristic example is two teen-age motorists heading for each other on the highway with their girl friends looking on, to see which motorist will first swerve aside. This game is a pure test of nerve, and the loser is Chicken. Schelling's second example is the air corridor to West Berlin, where Soviet planes flew through the air lanes after notifying the U.S. of training exercises on a certain day; and Cuba 1962 shows up pretty soon – "about as direct a challenge as one could expect" – along with various other Cold War crises. The difficulty is that in the teen-age heuristic example the only payoff is extrinsic – facing down the opponent – while in the mathematical game the payoff is entirely intrinsic. Schelling notes this difference (1966 p. 118 n. 7) but in the ensuing discussion about Cold War crises drops it again and talks of chicken games as 'tests of nerve' (which they are not, mathematically). Plainly Schelling's imagination is playing over Cold War crises misinterpreted as deliberate Soviet tests of U.S. nerve, in which the intrinsic stakes – West Berlin, Quemoy – are trivial and the main interest is in whether the U.S. will be Chicken. What would Khrushchev want with West Berlin, anyway? The issue is not West Berlin, it is whether the U.S. is Chicken. Schelling's discussion reinforces this ideological stance, focusing our attention on the 'deliberate challenge' interpretation of Soviet behavior and covering over the question of what intrinsic stakes there are for the S.U. There were in fact very great intrinsic stakes for the S.U. – in East Germany, not in West Berlin. In this example the ideology is located in the heuristics rather than the mathematics; indeed from Schelling's discussion his examples are Prisoner's Dilemma for the U.S., not Chicken.

A contrasting problematic appears in a series of models from about 1960; Morton Deutsch's trucking game (Leader), Boulding's variants of Richardson Process models (1962, Ch. 2, 3), Charles Osgood's 'GRIT' model (expanded Prisoner's Dilemma), and the 2 × 2 Prisoner's Dilemma that was taken up so enthusiastically by experimental psychologists in the 1960's. (The enthusiasm of the international relations people like Schelling for Chicken is matched by the enthusiasm of the social psychologists for Prisoner's Dilemma.) The problem formalized by these models is, how does one induce trust and co-operation in an ambiguous situation of common and conflicting interests? In each of these cases one of the players is assumed to be the initiator of trust-inducing moves, his objective being to teach the other to co-operate. The real-life problematic is clear: how can the U.S. induce the S.U. to abandon its dangerous conflict behavior and shift to peaceful co-operation? Here the S.U. is seen not as a bluffing bully out to conquer the world, but as a rather short-sighted suspicious competitor, in an international situation inherently conducive to mistrust and suspicion, who has to be reassured of our good will and gently taught to appreciate its own long-run interest in *détente*. In both problematics the U.S. is the actor and the S.U. is the problem to be dealt with, but in opposite ways.

Another example is the rash of interest in voting models in the last twenty years or so, models with single-peaked and double-peaked voters' preference curves, together with the tremendous concern over disqualifying Arrow's impossibility theorem (1951). What these models are telling their users is that voting is an important aspect of American politics, that voters' preferences are important in determining policy, and that it is terribly important to be sure that majority rule (or in Buchanan and Tullock's more conservative version, minority veto power) leads directly to maximizing the general welfare – which is what Arrow's theorem asserted to be unprovable without unduly strong assumptions about preference orderings.

How did we deal with the ideological component of our bargaining and decision models? Presumably one solution would be to collect a variety of models constructed from different ideological bases, to free us from imprisonment in a misleading ideological stance. We collected bargaining models all right, but there are not that many and they mostly are rather similar in what they focus on and what they ignore. The difference of attention between the Chicken people (hawks) and the Prisoner's Dilemma people (doves) is real but occurs within an agreement on larger issues; the two models both treat the U.S. as the actor and the S.U. as the problem. One simple device I used was to reverse the players, to make the U.S. the blackmailer or the short-sighted greedy grabber and the S.U. the hero – note how we adopted the designations U.S. and S.U. to emphasize this possibility – but this did not really shift the focus of attention; it merely showed how silly and misleading the models were. The basic solution was to construct new models, a cybernetic bargaining model incorporating information processing and feedback, and cybernetic decision models emphasizing the interdependence between the bargainers and the bargaining situation; at the start of bargaining both are indeterminate but they reciprocally determine each other during bargaining. We used the Ellsberg, Schelling etc. models to highlight what is missing from these models but present in the cases. The discrepancies between the models and the cases then provide the problematic which went into the new models. Are our ideologies involved in noticing these discrepancies? Of course.

I turn now to the question of objectivity. What conception of objectivity is implicit in the procedures we found ourselves following to overcome ideological distortion? In 1972 I reported that I had found three conceptions of objectivity-subjectivity implicit in scientific practice. A theory or description is objective if it (1) describes the way the world is when we are not interfering with it; (2) if it gives a complete, well-rounded description rather than a biased or one-sided account; (3) if it reflects the characteristics of its subjects back to them in such a way that they recognize themselves and become more aware of what they have been. There may be a fourth conception around but I have not yet found it.

The conception implicit in our historical work was plainly the second. We treated a history as a one-sided account, an account from a perspective, and tried to supplement it with accounts from other perspectives. The estimates of the participants were also from a perspective, and we tried to work out the perspectives and interpretations of each of the major participants in a crisis. We also were aware of our own biases, and tried to correct for them by discussion within the research group. Our aim was a case history that would satisfy all objections within the research group and also could take account of all earlier histories. This makes objectivity an ideal to be approached asymptotically.

We also noticed the same process of endless correction going on in the history of histories. The best example is the histories of the 1914 crisis. The histories and memoirs appearing just after the war were extremely biased, each trying to show that some other country was guilty of bringing on the war. After a decade of controversy and opening of archives, the Schmitt history (1930) could appear as comparatively objective. But Albertini, (1942) brought out important errors and deficiencies in Schmitt's history. Our researcher first pronounced Albertini's history to be the most objective, but we later found that Turner (1970) had added important corrections. Also Fritz Fischer (1961) has reopened the whole case from a new angle which has not yet been adequately worked through. So sixty years later we are still pursuing objectivity for 1914, though Albertini (1942) is a good approximation. What then can we say of our own studies of crises of only fifteen to twenty years ago? We would be lucky to come close to the objectivity of Schmitt (1930) and our studies look good only in contrast with the ludicrously one-sided earlier accounts, all of which try to prove the S.U. guilty (or the U.S., in the Soviet-East German accounts). These earlier accounts, incidentally, are the empirical basis for a good deal of the international relations theory of the 1960's, such as Young (1968). How much objectivity can we expect from a theory based on fairy tales?

I now criticize this conception of objectivity on the basis of arguments by Mannheim (1936). In terms of Mannheim's discussion (1936, pp. 150–152, 164ff) this conception of objectivity is part of bourgeois liberalism. The liberal view is that science achieves objectivity insofar as a full range of ideologies is represented among social scientists, so each can criticize the biases of the others and supplement their one-sided accounts. The result should be a truth that is broader and more impartial than that open to any single ideological position, and that would be an approximation to objective truth.

This conception of objectivity is good as far as it goes, especially in the study of history; a range of ideologies makes for a more judicious case history than that written from a single slant. Also discussion is easier if people are willing to admit that they have a point of view and that other points of view also exist. The biggest source of frustration in our discussions, other than the incompleteness of our data, was the insistence of some researcher that he had no bias, that he had looked directly and impartially at the facts, and that it is rude to even accuse a scientist of being biased. When he went on to insist that there was only one way to explain certain events, discussion became extremely difficult.

But the liberal program for science has certain inherent limitations which make the claim of gradually increasing objectivity a form of self-deception. Notice first the implicit ideal of an impartial truth and an autonomous science – a truth that does not favor any particular standpoint from the start because it is achieved by discussion among all standpoints. The assumption of autonomy, of relative separation of science from practical politics, is false both in our research project and in political science in general. A liberal work like Gamson and Modigliani's *Untangling the Cold War* (1971) is certainly more balanced and objective than a propaganda diatribe like Horelick and Rush, *Strategic Power and Soviet Foreign Policy* (1966) because of the former's ability to empathize with different points of view and the latter's complete inability to do so; but the practical aim of policy advice is essential to both works. The Gamson work implicitly appeals to the U.S. government to correct its mistaken foreign policy, thus treating the U.S. government – or at least the Democratic opposition, Henry Jackson, say – as an open-minded rational agent. Similarly our crisis bargaining research is a contribution to the art of "crisis management", advice to the U.S. government on how to win a crisis without war.

In organizational terms, our project developed several communication channels with present and former U.S. government officials interested in crisis management, including the Naval War College, without infringing on the autonomy of science; but if we had developed working relations with some radical organization, that would be politicizing the University, destroying our integrity as scientists, and inviting administrative measures.

The same practical implication occurs in the choice of subject-matter. Both the Gamson work and our research focus on crises and crisis management by politicians. They do not investigate where crises come from – colonial disputes before 1914, U.S. expansion into Iran before 1945, etc. – that's economics, but we are political scientists, not qualified to discuss economic history. They do not investigate the domestic influences on who gets to make foreign policy decisions – that's domestic politics, not international politics. That is, the conventional subject-matter specializations serve to reinforce the practical bias of treating international politics as advice to rational government actors. If some of us in the research group tried to transcend these barriers, a variety of devices came into play to bring us back to the point.

A further consequence of this unnoticed practical bias is that certain lines of theoretical inquiry are tacitly ruled out. The liberal is not as tolerant as he thinks he is. For example, the last section of our case study was supposed to be 'outcome and aftermath'. In my study of the Berlin crisis I argued, rather speculatively and superficially, that the aftermath of Berlin-Cuba-Vietnam was a dangerously militarized U.S. foreign policy establishment prone to escalation and aggression; that we were headed toward wars as a consequence; and that the best preventive was revolution. A sober topic. This irrelevant diversion of mine was greeted with amusement by some and stony, outraged silence by others; there was no discussion of it. The liberal claims to be studying pure science, but in practice he has difficulty conceiving of topics that do not connect eventually to some kind of government policy; he has no clear conception of revolution as practical politics.

A larger assumption of the liberal conception of objectivity is that sufficient viewpoints are already present among American social scientists, that is that American society is already sufficiently diverse and open to support an objective social science.

The implication seems to be that radicals who get caught up in the rational-discussion-among-all-viewpoints pretense are allowing themselves to be used for liberal policymaking and propaganda. Their presence proves the broad-mindedness of the liberals but their radical suggestions are ignored with amused tolerance. They soon find that the problems to be discussed are *liberal* problems, defined by the current subject-matter specializations, methods, and hypotheses, and their role is to be the extremist critic within the liberal spectrum. They are no longer radicals but extremists (revisionists, etc.), characters in someone else's dream. Their own radical problems and theories are ruled out as going beyond the boundaries of the field, among other reasons.

Is there a different recipe for objectivity, then? The suggestion of Lukacs

(1923) is that the radical viewpoint is itself unbiased, because it is the viewpoint of a potentially universal class, a viewpoint representing the whole of society. The trouble with this suggestion is that his universal class, the proletariat for itself, does not exist; it is an ideal type. Nor do we have a party anywhere that is related to this class in the manner Lukacs specifies (1923, Ch. 10). Consequently radicals who based their claim to objectivity on their proletarian standpoint are deceiving themselves, pretending to live in the future classless society; their thinking is utopian in Mannheim's sense. That is, they treat the future as though it were already present (which it is in part) and they neglect the obsolescent contradictory tendencies which are still influential in their thinking and in reality. A perfect example is the opening paragraph of Ch. 5 of Lukacs. An even more drastic chiliastic utopianism characterized some New Left politicians in 1968, as they participated in the final revolution that would sweep away all class, racial, and sexist oppression forever.

A more modest claim is that of Goldmann (1952, pp. 52–62), who argues that the Marxist ideology is potentially superior to others because it exposes the roots of all ideologies in social life. An ideology that points to its own social sources is potentially self-critical; it exposes its own limitations and distortions (p. 52). It forces people to recognize that their own thinking is limited and biased by social background and personality. Recognition of one's own biases may in the long run lead to some correction of bias; but in the short run it certainly leads to modesty. We, no more than liberals, can claim to be free of self-deception, wishful thinking, and one-sidedness in our research. We cannot even claim to correctly know our own biases, since self-deception can occur in investigating one's own limitations as easily as anywhere else.

Consequently we, like the liberals, need to organize multiple viewpoints for mutual criticism, using some of the diverse thinking and political experience available to achieve whatever objectivity our present society makes possible. There is no royal road to objective truth.

State University of New York at Buffalo

NOTE

* I wish to acknowledge the helpful criticisms of Alasdair MacIntyre and Andrew McLaughlin.

REFERENCES

Albertini, Luigi. 1966 (originally 1942). *The Origins of the War of 1914*. London: Oxford University Press.

Arrow, Kenneth. 1951. *Social Choice and Individual Values*. New York: Wiley.

Boulding, Kenneth. *Conflict and Defense*. New York: Harper.

Buchanan, James and Gordon Tullock. 1962. *The Calculus of Consent*. Ann Arbor: University of Michigan.

Diesing, Paul. 'Subjectivity and Objectivity in Social Science,' *Philosophy of the Social Sciences* **2** (1972), 147–166.

Dörnberg, Stefan. 1968. *Kurze Geschichte der DDR*. Berlin: Dietz.

von Eckardt, Felix. 1967. *Ein unordentliches Leben*. Wien: Econ.

Ellsberg, Daniel. 1968. *The Theory and Practice of Blackmail*. Santa Monica: RAND.

Fischer, Fritz. 1974. *World Power or Decline*. New York: Norton.

Gamson, William and A. Modigliani. 1971. *Untangling the Cold War*. Boston: Little, Brown.

Goldmann, Lucien. 1973 (originally 1952). *The Human Sciences and Philosophy*. London: Cape Editions.

Horelick, Arnold and Myron Rush. 1966. *Strategic Power and Soviet Foreign Policy*. Chicago: University of Chicago.

Lukacs, Georg. 1923. *Geschichte und Klassenbewusstsein*. Berlin: Malik. Reprinted 1967 by de Munter, Amsterdam.

Macmillan, Harold. 1972. *Pointing the Way*. London: Macmillan.

Mannheim, Karl. 1936. *Ideology and Utopia*. New York: Harcourt, Brace, Harvest Books.

Possony, Stefan. 1964. 'The Challenge of Crisis,' in Frank Meyer, ed., *What is Conservatism*? New York: Holt, Rinehart.

Possony, Stefan. 'Mao's Strategic Initiative of 1965 and the U.S. Response,' *Orbis* **9**, No. 1 (1967).

Salinger, Pierre. 1966. *With Kennedy*. New York: Doubleday.

Schelling, Thomas. 1960. *The Strategy of Conflict*. New York: Oxford University Press.

Schelling, Thomas. 1966. *Arms and Influence*. New Haven: Yale.

Schlesinger, Arthur. 1967. *A Thousand Days*. New York: Houghton Mifflin.

Schmitt, Bernadotte. 1930. *The Coming of the War*. New York: Scribner's.

Slusser, Robert. 1973. *The Berlin Crisis of 1961*. Baltimore: Johns Hopkins.

Smith, J. E. 1963. *The Defense of Berlin*. Baltimore: Johns Hopkins.

Steinbruner, John. 1974. *The Cybernetic Theory of Decision*. Princeton: Princeton University Press.

Turner, L. C. F. 1970. *Origins of the First World War*. New York: Norton.

Young, Oran. 1968. *The Politics of Force*. Princeton: Princeton University Press.

LEON J. GOLDSTEIN

TOWARD A LOGIC OF HISTORICAL CONSTITUTION*

I

The question which it is my intention to explore in the pages which follow is, What makes an historical reconstruction acceptable? But before I begin, it seems appropriate to say something about the term 'historical constitution,' particularly since so far as I know I am the only one who uses it. Such perverse idiosyncracy ought not to be indulged, yet in the present case some justification can be offered. I am not wedded to the term for its own sake, but it does seem to suit my purpose admirably. What I am trying to do when I use it, is to avoid using the established alternatives, because those alternatives carry along with them suggestions about history and historical knowing which I should want to reject.

Much philosophical effort has been expended in recent decades over the problem of historical explanation, but that will not concern us here. Prior to the effort of any historian to explain some historical event or course of events, must be the attempt to establish that something-or-other did in fact take place, that very something-or-other which will subsequently have to be explained. And it is at that prior stage that the epistemological problems of how the past is known arise. But whatever those problems are, and whatever solutions to them seem best, at that stage the historian is engaged in a variety of activities the goal of which is to make known what it is reasonable to believe happened at some time in the human past. It is by that set of activities that the historian constitutes the historical past, and the end-product of what he does during that stage of his work might be called the "constituted historical past" or the "historical past as constituted in historical research."

But why use such terminology when there seemingly exist perfectly ordinary ways of referring to the sort of activity in question? Thus, in the very opening sentence of this paper I use the common term 'historical reconstruction,' and one frequently sees the term 'historical description.' In my view, both of these expressions carry with them connotations which may tend to distort our understanding of historical knowing. And when one attends to how much of what passes for epistemological inquiry these

R. S. Cohen and M. W. Wartofsky (eds.), Epistemology, Methodology and the Social Sciences, 19–52.

days begins not with the ways quests for knowledge are actually carried on, but with what certain words, deemed relevant to whatever issue lies before the writer, mean, it becomes particularly pertinent to avoid using words which may have such an effect. It is perhaps not always possible to dispense with them entirely – as witness my opening sentence – but one function of the present discussion is to make explicit what they cannot mean should they from time to time appear in what I write.

Let us begin with the expression 'historical description.' In itself it is harmless enough. We could agree that historians' accounts of what it seems reasonable to believe took place in the human past might be called by that expression. Unfortunately, even if some of us were willing to allow the meaning of the expression to be exhausted by its extension, there are others for whom the overtones of the ordinary word 'description' would suggest what I take to be a mistaken conception of historical truth. Typically, a description is an account of something observed or witnessed. The truth or falsity of a description in that typical sense is determined by the extent to which it accords with what has actually been seen or witnessed. If one reads or hears a description which seems for whatever reasons implausible, one may sometimes even have the opportunity of testing its truth by setting out oneself to observe the object of the description. The main point here is that in this typical sense descriptions involve perceptions, and the truth of descriptions involve the concordance of what might be perceived with the account given of it. Of course, no one thinks that historians perceive the events they purport to describe. But many have no difficulty with thinking that historical descriptions must satisfy the same conditions of truth as do descriptions of the typical sort, that is, they must accord with what might be perceived by anyone present at the time who might be in position to witness the event. Yet this is a condition totally irrelevant to the practice of history. Historians – as distinct from witnesses present at the time – do make judgments of truth and falsity with respect to the accounts of past events proposed by their colleagues, and if philosophy of history has as one of its tasks the clarification of the bases for such judgments, then it wants to attend to what historians – as distinct from witnesses present at the time – do when they make and justify such judgments. There are some who would wish to distinguish between the grounds for truth of descriptions of all sorts, and the particular ways in which truth is established under varying circumstances. But that distinction is utterly gratuitous, for there is no way to show that historical descriptions satisfy the conception of truth in question. If we are interested in the actual practice of the discipline of history and the way in which historians-at-work make judgments of truth and falsity, we ought to attend to those intellectual – in no sense perceptual – activities in terms of which historians come to have some idea of what we may have reason to believe once transpired. And this makes it clear that the accounts they produce are not descriptions in the usual sense of that word. Put somewhat differently, if the way in which an account is put together – its genesis and function within some mode of inquiry – is relevant to the determination of its epistemic status, it obviously follows that descriptions of the typical sort are epistemically different from accounts produced by historians. To call the latter 'historical descriptions' without having attended to the considerations which have occupied us in the present paragraph is to risk being mistaken about the conditions of truth in the discipline of history as well as about the actual character of history as a discipline.

The term 'historical reconstruction' is perhaps less of a problem than 'historical description,' presumably because 'reconstruction' has a less obvious epistemological sense than 'descritpion.' The term suggests a putting together of what once was but is no longer, and although there are or were philosophers who took their task to be rational reconstructions of one sort or another, the basic sense of the term suggests an activity which is not intellectual. I do not actually know what the origins of the term 'historical reconstruction' are, and some may think that it is simply metaphorical. But it does seem to make sense if it arose in something like the following way. It may have to do with a not uncommon view of the way in which historians put togehter – literally reconstruct – historical events out of evidence. On this view, a classical statement of which may be found in the well-known handbook of Langlois and Seignobos,[1] the historian amasses his documentary evidence – the only kind of evidence discussed by Langlois and Seignobos – subjects it to critical scrutiny, and emerges with a collection of discrete affirmative assertions each one of which embodies an historical fact. This is followed by the synthetic activity of the historian whereby he determines which facts ought to be put together so as to produce a more or less complete account of the event in which he is interested. The historical event which is presented in his account is constructed or reconstructed out of the facts which emerge from the first or 'analytical' phase of his work. But this conception of Langlois and Seignobos is woefully out of date; indeed, at the time of its first publication it was already out of date by not less than three centuries.[2] If historical accounts could only be put together out of such antecedently established building-block facts as those required by the

Langlois-and-Seignobos conception of historical synthesis, very little of the already-established historical knowledge at our disposal would in fact exist.[3] This will become more apparent as we proceed. But for the moment, we may simply note that if the idea of an historical reconstructuion does implicate their idea of historical synthesis, the term is clearly misleading.

Having considered the two standard expressions for the activity which will be the focus of our attention in what follows, I could now actually use them freely with the expectation that those who read my remarks will be careful not to impute to me the suggestions associated with those expressions with which I do not wish to be associated. There is, however, no assurance of this. All too frequently, philosophical critics have preferred to find the sense of an opponent's view in what the words he uses really mean, rather than in what he uses them to say. The whole character of J. L. Austin's attack on A. J. Ayer's phenomenalism – in the former's *Sense and Sensibilia* – is of that sort. Thus, it would seem best to use what I take to be an appropriate term.

The word 'constitution' I borrow from the phenomenologists, but whether or not they would approve of what I do with it is not for me to say. The problem of constitution for phenomenology is the problem of how an object or the intersubjective world is put together by experiencing subjects.[4] It is this focus upon the problem of putting something together on the basis of materials of a sort unlike the finished product which seems to me to be strikingly relevant here. That the world of objects is put together out of materials unlike the objects themselves is, of course, not peculiar to phenomenology. Hume taught that objects were built up out of impressions, and Russell that they are logical constructions out of sense-data. Thus, whatever the particular contributions of phenomenology to these matters, when those who work within the philosophical boundaries of phenomenological philosophy talk about the constitution of an objective world out of subjective materials, they are dealing with a problem or a range or problems known to the rest of us as well. But those problems are not relevant here. All I wish to note is that this process of world-apprehension or world-making is called 'constitution' by the phenomenologists. Why does it seem to me a term worth importing into our present context?

The focus of the problem of constitution is the process whereby the subject constitutes a world. The way I have just formulated the matter may suggest that what interests those who work on the question is how the world is constituted by the idiosyncratic subject, and I have no doubt that no small part of the phenomenological literature on the question assumes that this is the case. But it need not be. In the view of Alfred Schutz, in its most fundamental givenness the world is intersubjective,[5] and it is hard to doubt that had he taken up the question of constitution he would have attempted to show how an intersubjective world is constituted by means of certain *social*-cognitive processes. It is important to note that the problem of constitution *need not be* rooted in subjectivism, since the position I intend to develop in the course of the present paper requires that a somewhat analogous distinction be made between the historian – the idiosyncratic subject who pursues whatever historical investigations he is moved to pursue – and the discipline of history – that organized body of knowledge and techniques for the acquiring of knowledge which keeps the historian from falling into the abyss of idiosyncratic subjectivism. In any event, certain things begin to emerge from all this, not least of which is the way I conceive of how to go about answering the question indicated in the very first sentence of this paper. I said there that the problem we are to deal with is what makes an historical 'reconstruction' acceptable. It is now clear that I shall be moving in the direction of attending to the character of the discipline in terms of which the historical past is known. The historical past is *constituted* by historians in the course of historical research. It might seem reasonable to say that whether or not the outcome of a piece of historical research is acceptable must depend on the extent to which it accords with what the part of the human past with which it deals was actually like. And that would point us again in the historical-description direction. But the difficulty with taking that direction is that it leads nowhere. There is no way to know that an historian's account *accords* in that way, because we know nothing about the human past other than what has emerged from historical research.[6] From this, it clearly follows that there is no way to test the adequacy of an historian's account against a past known independently of historians' accounts. The view of this paper is that the only possibility open to us at all for the justification of historians' accounts and the determination of their adequacy must be rooted in attention to them as the outcome of a certain epistemic procedure. One might think, with Dilthey and Marrou, of a *critique* of historical reason. But one must mean by that not the quest for immutable categories of the historical consciousness,[7] but rather the attempt to understand how the actually-practiced discipline of history is carried on, with particular reference to the character of making claims to knowledge – as well as criticizing and justifying such claims – within the discipline itself. It appears, then, that the whole of our present task may be located within a perspective which understands history as that disciplined way or set of methods whereby the historical past is constituted in historical research. Thus, my use of the term 'historical constitution' is not a linguistic idiosyncracy which ought to be replaced by the more usual terms discussed above. Rather, it is explicitly chosen because using it indicates something about my conception of the character of the objects of historical knowing as constituted in research, not as objectively there (*gegenständlich*) waiting to be discovered.

II

The task before us is to characterize that intellectual process out of which the historical past is constituted. But why do that in a paper which purports to be philosophical? Would it not be more appropriate to do that in a handbook of historical method, a work which takes as its task the instruction of the neophyte on how to be an historian? But the fact of the matter is that the handbooks never raise the question, much less point to an answer. What one typically finds in books of that sort are instructions on how to deal with historical evidence so as to derive therefrom the maximum amount of information. They tell us about the varieties of evidence,[8] though they tend to be preoccupied with written evidence, and offer recipes for deriving historical facts from it. Beyond that, they tend to be unsatisfactory and cast precious little light on how historians produce accounts of the human past on the basis of the evidence as it emerges from the applications of their critical techniques. According to the classic handbook of Langlois and Seignobos, the outcome of the application of critical analytical techniques to evidence is that the historian is left with a collection of facts – one fact embodied in each of the simple declarative sentences of his documentary evidence – and from this collection he chooses the facts he needs in order to produce the account he wishes to produce. This, as I shall try to indicate shortly, is entirely false to the procedures of historical constitution it purports to describe. Marrou has a strong feeling that what Langlois and Seignobos say about the intellectual activity we are concerned about here cannot be correct, and is most unambiguous in his belief that what is needed is an account of the historian's mental or intellectual processes in order that we may understand how he actually moves from his evidence to the events he purports to present to his readers, but he does not seem to be able to say precisely and in detail what such an account would be like. If we are to have rules for becoming an historian, I suppose they ought to be produced by people who are themselves historians. But, so far as I have ever managed to ascertain, no such rules have been formulated. The handbooks tell the beginner how he is to handle his evidence, but they tell him exceedingly little about how he is to derive the historical past from it.

But if we have no account of how to derive the historical past from historical evidence, we do have all manner of writing in which that is done. Thus, we do have a body of claims to knowledge which may provide occasion for philosophical scrutiny. Our concern, then, is not to produce a how-to-do-it account to serve in place of those the handbooks fail to provide, but to attend to the character of the intellectual procedures implicit in the production of the kinds of claim to knowledge which are the outcome of historical investigations. If handbooks actually contained explicit instructions for deriving the historical past from historical evidence, the answer to the question this paper attempts to deal with would be comparatively easy to arrive at. One would point to the way in which some historical account was derived in accordance with the rules and could discern with relative ease why it was – or was not – an acceptable 'reconstruction.' But instead of having so easy a path, it will be necessary for us to look ourselves for the direction in which the answers lie. For this, a look at some simple examples of historical constitution may be of some help.

Let me begin with the simplest example I can think of. In Collingwood's 'Roman Britain'[9] we are told that in fifth-century Silchester "a tombstone was found ... written in the Irish, as distinct from the British, form of Celtic." From this bit of evidence, Collingwood derives – which is *not* to say deduces logically – the following historical fact:[10] "An Irishman who died in Silchester and left friends able to make him an epitaph in his own language must have been a member of an Irish colony in the town."[11] Collingwood's conclusion seems entirely reasonable, but that does not make it correct. Controversies in history are frequently conflicts over what it is reasonable to believe in view of the evidence and *the ways in which historians think about evidence*. Such conflicts are apt to be the opposition of what seems entirely reasonable to what seems entirely reasonable, thus not everything which seems reasonable will prove to be historically true. To speak of something as seeming reasonable may seem to carry with it the suggestion of subjectivism: it seems reasonable to me, but it might not to you. But the whole point of our inquiry is to find an answer to the question of why an historical account seems reasonable to believe, thus what we want to have are reasons which move us beyond the merely subjective to the sorts of consideration which would be cited in rational defense of a claim to knowledge. With respect to the example from Collingwood, it is not clear to me why it seems reasonable, but we may note from the start that two things which might make the assertion of an historical fact seem reasonable may not be said of it. One of them I have already noted in passing: while Collingwood may be presumed to derive his historical fact from the evidence, the derivation is not an instance of logical deduction. I shall elaborate somewhat on this point in the immediate sequel. But first, I would like to say what the second thing is, namely it cannot be claimed that Collingwood read the fact out of the evidence. Our evidence contains no sentences informing us of the existence of such a colony. Collingwood asserts that it existed only because the existence of the tombstone becomes intelligible if there were such a colony. Thus, reflecting on his evidence leads the historian to propound the existence of something for which he has no direct documentary evidence. It is here that we find at work what Collingwood himself thinks of as the historical imagination, and Marrou and others as historical reason.

I want now to pick up the point about logical inference. It may seem obvious to some that with the critical establishment of the evidence, the historical facts ought to be derived logically from it, and one even finds the relation of historical facts to historical evidence formulated in that way.[12] It is not necessary for us to introduce a formal symbolism in order for it to be seen clearly that Collingwood's statement about the Irish colony does not follow deductively from the one about the tombstone. But some may wish to say that it must presuppose some statement or other which taken together with that about the tombstone would in fact make it possible to derive deductively the one about the colony. I do not wish to take this matter up in detail, but I do wish to indicate something about why it is a most implausible view. To begin with, it is a point of view altogether wooden in its application: any conflict between historians as to what the evidence makes it reasonable to believe would immediately be construed to be an implicit conflict over what is presupposed. Those fields of historical inquiry which are much worked on and in which unsettled questions are reflected in much controversy would find themselves haunted by all manner of ghostly assumptions. In addition, if we are interested in knowing something about the character of the thinking which results in the constitution of historical events, we are not helped by being told that any constitution rests on the sort of assumptions that would, in logical fact, make it possible to derive the conclusion deductively from the evidence. Since Collingwood has not derived his statement about the colony from the evidence deductively, we want to know why it seems reasonable. To say that a deduction is possible in spite of what the historian did, is merely to insist upon some point of principle and, at the same time, to display a lack of interest in what the character of historical constitution is actually like.

I have, in effect, said two things about the relations of the past fact constituted by Collingwood and the evidence upon which the constitution is based, namely, that *he did not derive* the former from the latter deductively and that the relation between the two – however he actually derived the one from the other – *is not* deductive. There is, however, more that may be said about this issue, and while I do not think that this is the place to deal with it further in detail, I do wish to indicate something about what sorts of consideration are involved. Basically, what is involved are the two inter-twined themes of the evidence relevant to some historical problem and the character of disagreement with respect to historical constitution. With respect to the first of these, while Collingwood's account makes it clear that his belief in the colony depends on the existence of the tombstone and that without the tombstone he would not have mentioned – perhaps even thought about – the possibility of such a colony, there is really more to it than that. Collingwood knows not only about the tombstone but about a host of other evidence which go into the constitution of the history of fifth-century Britain. Thus, it might be argued that Collingwood's statement about the colony does indeed have a deductive relation to the evidence, but the evidence is to be construed rather more broadly than simply the tombstone. But even if there were merit to this suggestion, it must again be noted that the thrust of such considerations would be in the direction of a post-constitution justificationary reconstruction of the historian's activity along the lines of certain preferred logical principles. This would be of no help at all if we are interested in the actual character of historical constitution as the intellectual activity in terms of which the historical past becomes known.

But the idea of relevant evidence neatly held together in virtue of being the evidence relevant to the constitution of the same bit of the historical past is altogether too abstract and very far from the actuality of historical practice. If one attends in detail to the character of controversy in some field of history in which the basic issues still concern the way in which the past is to be constituted, one discovers quite early on that one central feature of the controversy is disagreement over what the relevant collection of evidence is which is germane to the problem. Part of the point at issue is to determine which documents, for example, belong with what other documents, artifacts, ruins, and the like, as conjointly being the evidence for the same narrowly-construed swath of the historical past. In the first decade of research instituted by the discovery of the Dead Sea Scrolls, when scholars were far from agreement as to the general character of the texts and the community which produced them, and when arguments were being put forward linking the scrolls to such diverse groups as the Essenes, the Holy Congregation of the very early Rabbinic tradition, and the Sicarii of the Jewish revolt against Rome nineteen centuries ago, one finds that different scholars associate the scrolls with different bodies of surviving ancient evidence, and where they may sometimes refer to the same evidence, they will use it differently in different ways for different purposes. Nor have we to do with an arbitrary, subjectively-tinged kind of activity. It is not the case that each historian piles up his evidence neatly and in order and then proceeds to the task of constitution. Were that the case, the logical-deduction view might seem to be plausible, differences among historians being understood as owing to the fact that the different piles of evidence have different logical consequences. The decision to order the evidence in one way and to re-order it in another is part of what happens in the course of the historian's attempt at constituting that part of the past on which he works. As he proceeds, he comes to think that evidence which antecedently seemed to have no relation to other evidence is, indeed, directly pertinent to it, and things he thought belonged together might not. It depends on how satisfactorily it seems to him the results of what he believed to begin with are turning out to be. As he moves along certain lines, it occurs to him that what he is doing may be furthered by what he once discovered in old documents, or what he has been learning about the past illuminates documents he had no antecedent reason to think had any relation to the matters of his inquiry. There is no point here to pursuing this further. The point to be made is that the logical-deduction point of view seems no more to make sense of the historian's activity of historical constitution in the typical complex cases than it did in the simple case of the tombstone and the Irish colony.

III

Having thus dismissed two views which may have seemed antecedently plausible, namely, that historical facts are read out of the evidence once the latter has been treated critically, and that the historical past is constituted deductively by logical inference from the evidence, what have we left? It is exceedingly difficult to answer this without further attention to what the actuality of historical constitution is. In addition, such attention may well be required for those for whom the initial plausibility of the views dismissed in my previous discussion has not been entirely removed. There is, after all, a certain abstractness to that discussion: it is rather more argumentative than illustrative and, consequently, cannot *show* that historical constitution is like this and not like that. Presumably, the way to correct this would be to become immersed in the writings of historians, but it must be noted at once that not all kinds of such writing would be of service to us. A good deal of such writing does nothing more than present historians' accounts of what various parts of the human past were like. My "nothing more" is not intended as denigration. It is only intended to indicate that there is something else, and it is that something else which concerns us here. Those accounts which most of us who are not historians usually read when we read history are the historians' reports to date of what they take the human past to have been like. They are the finished products of historical constitution, but they do not typically present us with the way in which that constitution was carried out. In a well-worked-on field, the historian typically begins with what the scholarly tradition has established, and on the basis of new thinking and new evidence produces such modification as he takes to be required and supportable. He presents, not infrequently in narrative form, his account of what the past was like, and for the most part that account is not interrupted in order to show how historical thinking about historical evidence led him to the conclusions he presents. He will, of course, make reference to evidence, but this will typically be informative only to his colleagues who know how the evidence is preserved and published. For the rest of us, it will provide only the assurance that the work is one of responsible scholarship, not an arbitrary figment of the historian's fantasies.

Thus, in order to get some idea of the nature of historical constitution, it is usually useful to examine instances of little-worked-on periods. In dealing with these, the historian is rarely far from his evidence – in some instances the non-historian reader gets a stronger dose of evidence than he can possibly want – and it is there that we might hope to see just how the conceptions of the past which are written up in history books are actually formed. In what immediately follows, I shall present a sketch of an example of this. Even if it should prove difficult to extract from the sketch an answer to the question with which this paper was begun, it should make it clear that the cognitive activity which is historical reconstruction is rather more complex than the views dismissed above would have it.

The point of departure for the inquiry we are about to sketch was a stone inscribed in runes found encased in the roots of a tree in Kensington, Minnesota, late in the nineteenth century.[13] The first response of scholars was to reject the inscription as a forgery. What would such a medieval text be doing in the heart of North America, and was it not an unusual coincidence that a Scandinavian text is purported to be discovered precisely in that part of the United States in which there was a heavy settlement of immigrants from Scandinavia? One may note, then, that the very first question which seems to arise here is whether or not what purports to be historical evidence for something medieval is, in fact, historical evidence for something medieval. The historian whose work we are considering, H. R. Holand, deals with the problems raised in ways which yield what seem to him to be satisfactory and acceptable results: Given the kind of tree the roots of which encased the stone, it could not have begun to grow before a certain time, which time was before the arrival of the immigrants from Scandinavia. Those who were bothered about the character of the language in which the inscription was made tended to compare it with the sagas of the eleventh century, whereas the inscription itself contains a fourteenth century date, and anyway, a modern forger who was intent on fooling the scholarly world – for whatever hard-to-fathom motive – would be equipped with the latest grammars and dictionaries and would produce a text the scholars would have no difficulty about accepting. Thus, the very defects of the text argue in favor of its genuineness. But we are still left with the problem of how to account for a runic inscription in the heart of North America almost a century and a half before Columbus is supposed to have discovered the place.

Of course, one does not simply account for an inscription. One has to account for an inscription having certain features, and among those are the things it asserts. This one talks about an expedition seemingly of both Swedes (Goths) and Norwegians and a bloody massacre, presumably at the hands of Indians. It mentions ships and it tells of things being determinate distances from other things, which distances are reported in a unit – "days-journey" – the length of which is not familiar. It may be noted that one of the problems that scholars found with accepting this text as authentic is precisely that they could find no instances of Swedes and Norwegians participating together in a joint expedition. Holand thinks that this last is once again owing to their tendency to compare the text with those which have come down to us from the eleventh century, but if one takes seriously the claim the text itself makes to be from the fourteenth century, there is, in fact, one known instance of an expedition in which representatives of both of these Scandinavian peoples took part. And that was an expedition under the leadership of Paul Knutson which was supposed to have set forth in 1355 at the behest of King Magnus Eriksson for the purpose of rendering aid to settlements in Iceland which were then beset by Eskimos, and which were, in consequence, breaking up with the settlers lapsing into paganism. Given that the Kensington Stone – which is what the runic text we are dealing with is called – is dated 1362, and taking into account the character of the sailing vessels in use at that time and the distances which would need to be traversed, to find remnants of an expedition from Scandinavia arriving in Minnesota seven years later is not the least bit implausible. There are other things stated in the inscription that Holand must look into. There is, for example, reference to a camp at a lake[14] having two skerries located one day's journey from the stone on which the text is inscribed. Holand scours the lakes within what he thinks it reasonable to believe a day's journey is, does, indeed, find one with two skerries and discoveres within it or at its shore evidence that might be the remains of an early encampment of the sort involved. Step by step, Holand constitutes an expedition that began in Northern Europe in 1355, passed along one or another plausible route, and arrived in Minnesota in 1362. The reasonableness of the account is further supported, or so Holand argues, by the discovery in the region of tools and weapons which, it is claimed, rather more resemble the sort produced in Scandinavia than that manufactured by the Indians of the Minnesota area. There are other matters discussed in Holand's account,[15] but I think I have presented enough to go on to some general observations.

When Holand's work is completed, we know something about an expedition we never knew before, about a complex historical event never before described, never recorded in our historical sources. Holand's account is not made up out of statements found in the evidence. We did, apparently, know of an expedition that was thought to have left Scandinavia for Iceland in 1355 even before Holand produced his book, but we know precious little about it and about what befell it in the course of its travels, if it ever actually started out. I am in no position to say that Holand's account is the last word on the subject or even that it is the most acceptable account, given what is known. But I do want to say that it is an attempt to show why it is reasonable to believe something about the human past which no one, presumably, thought about prior to this effort at historical constitution.[16] We have before us the outcome of Holand's constituting activity,[17] but is there anything we can say about what it consists in? And is it possible to say why it might seem like a reasonable, even acceptable piece of work?

Before we try to consider what might be said about this, I should like to make what might be considered to be a relevant digression. It is relevant in that it does deal with a matter of the historian's procedure, and, in fact, clarifies by reference to the specific example of Holand's work a point discussed somewhat more abstractly above. Yet for all that, it is somewhat digressive in that it does not deal immediately with the questions propounded at the end of the immediately preceding paragraph. In any event, I did argue above against the idea that prior to the historian's activity of constituting the historical past he had first to put together the evidence relevant to his problem and then proceed with the job in hand. I said there, that part of what emerges from the course of historical work is a view of what the evidence is that belongs together as relevant to the solution of some particular historical problem. There is no way to tell from the evidence itself that it belongs naturally with some other evidence and must obviously be used together with it for the constitution of the same historical event or course of events.[18] Allowing that the Kensington Stone is an absolutely essential piece of evidence for the inquiry which Holand undertook, on the ground that without it would not have occurred to anyone to think of the possibility that Norwegians and Swedes had reached Minnesota in the fourteenth century, can we say that it has some obvious connection with the evidence in terms of which historians constitute the beginning of the Paul Knutson expedition, and that the stone and the other evidence clearly – and antecedently – belong together as evidence for the same course of events? This hardly seems to be a reasonable view. To begin with, at the very moment of the discovery of the Kensington inscription what was most reasonable was to suspect its authenticity – somewhat analogous to the way in which it was at least not unreasonable to be dubious about the results of the Michelson-Morley experiments when they were first reported – and this would make no sense if its connection with the Paul Knutson evidence was obvious and natural. It is only when it becomes clear to Holand that the arguments against the inscription's authenticity cannot be sustained, and that the question of how a stone with a runic inscription became encased in the roots of a tree in Minnesota becomes a serious one, that the possibility of finding a connection between the two sets[19] of evidence becomes something to consider. There is, of course, nothing necessary about even this. An inscription dated 1362 and discovered in 1898, has more than five centuries to be moved about, and with sufficient freedom of imagination I suppose one could come up with all manner of hypotheses as to how it got to where it was found. Still, once the historian is satisfied as to its authenticity, it is not unreasonable to think that it may well have been inscribed in the area in which it was found, particularly if that is an area of relative wilderness rather than one over which much civilization has passed. Having reached that stage – accompanied by the obvious recognition that if Norwegians and Swedes were so far from home in the middle of the fourteenth century there must surely have been some not inconsequential expedition which resulted in their being there – the historian discovers that indeed there was a recorded expedition of the sort in question, the very first in recorded history in which both Swedes and Norwegians collaborated. It is only at this point that it makes sense to entertain the possibility that the two kinds of evidence may be brought to bear on the same subject.

IV

The pieces fall into place. A motley of odds and ends are made to fit together so that a reasonably coherent picture emerges. What starts out as implausible, namely, that there could be authentic, fourteenth-century Scandinavian remains in Minnesota that were not brought there at a much later date, turns out to be not all that implausible after all. In one sense, what we are looking for here is an explication of the concept of historical plausibility, and whatever it may turn out to mean, I suppose it is clear that it is not a logical idea in the formal sense. An acceptable historical constitution must present us with an account which is plausible, and that seems to mean one that it is reasonable to believe given what we know. As long as there was no reason to believe that Europeans had reached the Americas before the voyage of Columbus, the existence of the Kensington Stone was certainly perplexing. Even when it became evident that Scandinavian explorers and settlers had appeared rather early on the northeastern shores of North America, Minnesota is surely a long way from there.

In any event, we have some notions which seem relevant to the matter which has been our concern throughout the present paper, yet if one looks at them closely, one cannot really claim to have characterized historical constitution. One may say that in order to be acceptable, an historical constitution has to make something historically plausible. And one may say that historical constitution is the result of thinking historically about historical evidence. But we seem not to know how to explicate historical plausibility except in terms of the idea of seeming reasonable to believe. And we do not know what to say about historical thinking about historical evidence other than that it is not deductive thinking.

One might suppose that if the relation between historical evidence and historical events as constituted in historical research is not deductive, it must be inductive. It is difficult to speak with certainty about a subject on which the specialists themselves, the logicians, do not agree and concerning which there seems to be so much argument. Yet my strongest suspicion about the matter is that the relation between the two cannot be inductive. The logicians tell us that in inductive reasoning we widen the scope of our knowledge inasmuch as inductive conclusions take us beyond what is asserted – explicitly or implicitly – in the premises of an argument. Inasmuch as it is plainly the case that historical accounts say considerably more than what is contained in the evidence which is the historian's point of departure, it might seem that historical thinking is precisely a form of inductive reasoning. And when one attends to the variety of different things which are discussed under the rubric 'induction,'[20] it might not seem all that troublesome to add historical thinking to the list. But there are at least two reasons for not doing this. Whatever induction is, logicians speak of it as a kind of inference: Barker[21] takes it to be a mode of 'nondemonstrative' argument or 'nondemonstrative inference' and Salmon, in the work just referred to, tries to show in the case of every one of the different sorts of induction he deals with how it looks as a form of inference having certain logical characteristics. In addition, the kinds of matter which is typically discussed when logicians are talking about induction, the kinds of example introduced to make clear what they have in mind, indicate that inductive inference involves classes of things having properties of sorts, making predictions about the characteristics of populations on the basis of partial samples, and such like. Neither one of these may be said to characterize the sort of thinking which goes into historical constitution.[22] It is exceedingly hard to discern a line of inference in either of the examples that we have attended to in this paper. What is the inferential path that takes Collingwood from the tombstone in Silchester to his conclusion with respect to the Irish colony? And how can one ever find the direction of inference that takes Holand from the Kensington stone, and the other evidence which the course of his investigation proves to be relevant to the work he is doing, to the final result which is the constitution of an historical course which begins with the Paul Knutson expedition and ends with the massacre of part of its remnant and the inscribing of the runic text in Minnesota? Nor is it possible to say that historical thinking is like arguing from the character of a sample to that of a whole. It makes no sense to say that back of Collingwood's conclusion lie correlations of the sort discussed by logicians when they deal with induction.

Historical-evidence-in-hand is in no way like a sample of a larger population and is not treated as if it were.

It one wants to identify historical constitution as a specimen of a determinate kind of logical thinking, we could say, following Peirce, that it is a kind of abduction, but in the end this will not prove to be clarifying. There are a number of discussions of abduction in Peirce's *Collected Papers*, and I think that the following characterizations of it are typical of what he says. In one place he tells us that abduction "consists in examining a mass of facts and in allowing these facts to suggest a theory. In this way we gain new ideas; but there is no force in the reasoning" (8.209). Elsewhere, he says "Deduction proves that something *must* be; Induction shows that something *actually is* operative; Abduction merely suggests that something *may be*" (5.171; italics in original). It is abduction, not induction,[23] which in Peirce's view is the source of new ideas which fertilize science and give it new direction, and he goes so far as to say that "every single item of scientific theory which stands established today has been due to Abduction" (5.172). Finally, Peirce claims that "Anything which gives a rule to Abduction . . . puts a limit upon admissible hypotheses" (5.196), and this seems to suggest that we cannot expect to be able to offer a formal characterization of what abduction is inasmuch as such a characterization would presume to put some limit on the application of intelligence and imagination to the data which set the problem to be dealt with.[24]

Even our meager sample of two instances of historical constitution makes it clear that to think of historical thinking as a sort of abduction does not seem unreasonable. Here, too, what emerges from the course of the historian's investigation results from thinking historically about the data before him in order to emerge with a conception of the human past which is richer by far than that data. What thinking about those data suggests is only what may have been, surely not what must have been, there being, as he says "no force in the reasoning." If historical thinking is, indeed, a form of abduction, then here, too, one has to say about our knowledge of the historical past what Peirce says about scientific thinking, namely, that all that we have of it is owing to this way of thinking. And then we are forced to face the last of the passages quoted, that abduction cannot be limited by a rule. This means that being told that historical constitution is a form of abduction is not to be told anything about the actual character of this mode of thinking – if it is only one mode of thinking. We are, in effect, left with the question with which this paper was begun, What makes an historical reconstruction acceptable? rather like Peirce's own "What is good abduction?" (5.197).

Not infrequently when seeking to deal with abduction, which he sometimes called "retroduction," Peirce would refer to or even discuss Kepler's work on the orbit of Mars, which work he highly esteemed (cf. 1.71 ff.), but calling it abduction or the "greatest piece of Retroductive reasoning ever performed" (1.74) doesn't actually leave us with an idea of its nature as being told Euclid's *Geometry* is deductive in character or that Spinoza's *Ethics* attempts to use deductive techniques to establish metaphysical conclusions does about those works. In order to know just how Kepler reached his conclusions, N. R. Hanson was required to devote a large part of a substantial work to examining it in detail.[25] But what he emerges with is certainly not the logical form of abductive inference to which any other instance of abductive reasoning – including historical constitution if in fact it is abductive reasoning – might be expected to conform.

V

The outcome of an instance of historical constitution is a claim to knowledge which the claimant – and typically others as well – takes to be reasonable, and we have still to determine what being reasonable can mean here. I think that it must be clear by now that the reason that this is a perplexing question is owing to the increasingly apparent fact that between the claim to knowledge – the historian's constituted event – and the data, the historical evidence which leads to his constituting activity, there is what might be called an epistemic gap. I think it may be said in general that whenever abduction or abduction-like activity is called for it is because there is no direct way to bridge a starting point and a conclusion. If the relation between historical evidence and historical events were a deductive one, clearly there would be no gap. Nor would there be a gap – or, at any event, so large a gap – if any sense could be given to the idea that historical events resemble historical evidence. Were we to have half a dozen or so portraits of someone, say, from the eighteenth century, that is, before the availability of techniques for photographic reproduction, and if the portraits resembled one another, we could presumably believe that the subject of the portrait resembled his portraits in the way that they resemble one another. We would thus conclude that we have some *idea* as to how he looked. Quite obviously, there is no such resemblance between our *idea* of some historical event and the evidence on which that idea is based, though it may be that those who tend to think of history as drawn from documents are trying to think of something analogous to that.

Documents, of course, are not irrelevant. But they do not contain the complex and continuing events which emerge from the historian's act of constitution. Nor does it make sense to think of those complex accounts which we find in books of history as generalizations, as it were, of what is found stated as particulars in the documentary evidence. A. M. MacIver may be suggesting such a relation between the two when he speaks of contrasting "the 'generality' of historical statements with the individuality of the facts on which they are based."[26] He goes on to say that an historian's report to the effect that "The Normans defeated the English at Hastings in 1066" is the sort of historian's generalization he had in mind, and that what it generalizes is all of the specific and detailed facts which go to make up the battle, its course and its outcome. The unfortunate suggestion of MacIver's claim is that there is some way in which those particular facts actually figure in the historian's work, whereas the actuality of the situation is that he does not know them at all. He has reason to believe that the battle took place and the outcome was what is asserted, but whatever that reason is it is not because he knows who shot whom, and so on. What MacIver has done is confuse the explication of the concept of battle with how an historian knows that some particular battle took place.

The example of the Battle of Hastings, however, is not really helpful to us here. It would not be the least surprising to learn that there are actually extant documents which make reference to that battle, and anyone who takes seriously the view that the historical facts are dug out of the documents would likely to be able to use that example to make his point. So it is best to drop it, taking note only of the fact that contained in the way that MacIver makes use of his example is a hint of a sound point. And that is that the historical statement is related to data of a sort unlike itself. It is not, of course, related as a generalization – in his special sense – to specific factual data. One may suspect that he had a glimmer of something, a sound intuition which was not made clear and which, in consequence, he expressed incorrectly in terms of a poorly chosen example.

More helpful will be the following observation by Marrou:[27]

In his thesis,[28] J. Schneider studied the development of an original social class, the patriarchal society of Metz. He based his findings partly on a very carefully constructed, criticized, interpreted and exploited file of some two hundred legal documents concerning financial operations, principally real estate, carried out by the bourgeois of Metz between 1219 and 1324. It would be wrong to imagine that those elementary "facts" are more concrete, more real and more historical than the phenomenon of the collective fact – the transforming of an urban oligarchy into a landholding aristocracy. But in a still more obvious way than in the case of small individual facts, the role of the operative procedures selected by the historian appear to have a decisive role in determining these 'facts' of a global type.

Marrou then goes on to discuss the problem of determining rates of variation in mortality for any given region of rural France in the seventeenth and eighteenth centuries, the accumulation of 'elementary "fact" ' – presumably corresponding to the "concrete" facts of the quotation above – and the determination of techniques appropriate to the extraction of 'global' facts – the 'collective' facts of the quotation – from the elementary facts.[29]

What sets Marrou's problem is that the global fact is not contained in the elementary facts individually or taken together. If one follows the procedures of Langlois and Seignobos and other handbook methodologists who think along their lines, and sets out to extract from each of the two hundred odd legal documents referred to in the long quotation from Marrou all of the declarative sentences it contains, subjects them to the most critical examination, and emerges from the enterprise with a collection of unimpeachable assertions each of which is said to embody a fact, or a concrete fact,[30] one still does not have the collective or global fact of the transformation in medieval Metz constituted in Schneider's book. Nor would there be any way of ordering the concrete facts so as to alter this state of affairs.

Marrou is terribly sensitive to this situation and it is the main point of his book to deal with it. Something happens when these concrete facts are subjected to the scrutiny of historians, and that something leads to the emergence – constitution – of collective facts. The passage quoted above makes reference to historians' "operative procedures," and the second chapter of his book is called 'History and the Historian are Inseparable.' Marrou knows that the only way we are going to be able to understand how historical events are constituted is to attend to those 'operative procedures' in ways that the handbooks seem never able to do. One may note that Marrou's inclinations seem somewhat Kantian, and the reader expects that at anytime he will attempt to provide a transcendental deduction of the categories of historical reason. Even if one has not much sympathy for that sort of conception of the problem and doubts if it is the least bit feasible, one must at least recognize that Marrou has a better sense of what the problem is and where the solution lies than do the authors of typical handbooks. I should say myself that there is no way in which the quest for categories of historical reason can be successful,[31] but there is surely some relevant sense of historians' 'operative procedures' which is what we want to consider. In other terms, it is the elements which enter into the historian's abductive or abduction-like procedures which must be examined.

It should not be necessary to insist that our interest is not in anything that might be called the thought processes of historians understood in some psychological sense. It is by no means obvious that there are such special thought processes, and our concern here is with the sorts of things which enter into historical thinking. Our focus is on the work historians do – work which is always intellectual and consequently always involves thought – the thought which produces it and the things thought which emerge from it, and that only. There is no need to conjure up special thought processes in order to deal with historical constitution, but it is usually a sound precaution to try to make that clear since it is virtually inevitable that someone will try to foist such a reading on what I am trying to do here. What must be attended to are the kinds of intellectual or theoretical or conceptual conditions which enter into the constitution of the historical past.

VI

We have seen that historical constitution is an abduction-like mode of thinking in which the historian thinks about historical evidence and emerges with a conception of what some part of the historical past was like. Obviously, the account which emerges must make sense of the evidence, but beyond that it says considerably more. We must now try to come to some comprehension of the sorts of thing which enter into the constituted account.

To begin with, we might speculate about what sorts of consideration might have entered into the writing of history at a time prior to the development of the systematic, social-scientific study of human affairs. It is hard to doubt that for an historian of such a time what is plausible or not is determined by the sophisticated common sense of his time and place. Readers of Herodotus know how frequently, after reporting some – for him – hard-to-believe account, he observes that what he reports is what he was told and that he himself accepts no responsibility for its truth or falsity. Herodotus's own implicit doubts about the truth of these things is clearly rooted in the fact that no educated Greek of his time could lend credence to them. They could not be real, and no account of what purports to be a piece of the historical past could include them. When Herodotus tells us about those things he is not really working as an historian. He is simply telling us interesting things he has learned about the Egyptians, the Cimmarians and others in the course of his travels. If Herodotus could actually have been the historian of any of those peoples in the sense of constituting the course of their affairs in generations earlier than his own,[32] he would have to take rather a stronger stand with respect to those reports than he actually allows himself to take when he is only repeating what he had been told. In that case, even if his documents contained reports of those happenings, those reports would have to be taken as evidence for the possible beliefs of people but not for what actually took place. In sum, what is reasonable to believe is constrained by two factors, the evidence and common sense, and the imperatives of neither may be gainsaid.

In recent times, there has been a great deal of effort expended on psychological and social scientific studies. While it is not easy to determine the extent to which these researches have affected what most people have come to think about what is possible in human affairs, it is hard to doubt that the community of scholars among whom historians must be included has been affected and influenced. A point made by John W. Hall, in a report of a conference dealing with a certain period of Japanese history, seems to be relevant here: "Significant reinterpretations of major portions of a nation's history occur periodically as historians combine new findings with new approaches or as they pursue their inquiry with markedly different conceptions of the nature of the historical process."[33] That new findings, things which may function as evidence, affect history's conception of an age is rather a truism, and we need not be concerned with it here. It is the "new approaches" and "different conceptions" that are relevant to our interest at the moment, for it is these which affect the ways in which historians think about what the evidence suggests and what can be admitted into their accounts as plausible. In effect, what this means for the practice of history during the time since the emergence of the social sciences as systematic ways of coming to grips with the nature of social reality and possibility, is precisely that the sophisticated common sense in terms of which matters were settled by Herodotus is no longer the key to determining what is or is not acceptable in an historical account.

Actually, these last words require some qualification. One cannot claim that all historians take the findings of the social sciences and psychology into consideration when they set out to write their accounts of whatever swath of the human past they are writing about. Thus, it is no doubt the case that any number of presently practicing historians are still affected by their own sophisticated common sense of what is socially and humanly plausible. But that sophisticated common sense is rather markedly unlike that of Herodotus and the educated Greeks of his day, for no small part of what makes it up – no small part of what informs it – comes from the social sciences and psychology. An historian need not be explicitly committed to the utilization of those disciplines in doing his work for him to be affected by the ideas which those disciplines are disseminating into the intellectual atmosphere. Examination of the works of individual historians not particularly associated with the endeavor to make history scientific by means of the explicit effort to introduce into it the methods and theories of the social sciences – not, that is to say, people like Lee Benson, Robert Fogel or Charles Tilly – would presumably make it possible to discover just what some of the elements are. Unfortunately, that cannot be done here. But it does not seem unreasonable to believe that the common sense of educated people in the twentieth century has been importantly affected by the results of work in the social and psychological disciplines. It is entirely like the way in which our common sense with respect to the natural world is affected by the results of modern science even if we are not scientists and do not have much actual knowledge of natural science.

Some indications of the uses of social-science theory in the constitution of the historical past have been given by Robert Berkhofer in his well-known study of the relevance of such matters to the historian's tasks altogether.[34] Interestingly, he does it in such a way as to point to the distinction I made earlier between the descriptions of eyewitnesses and historians had his interests led him to want to draw that distinction explicitly. For what he does is contrast the situation in which social scientists – who as observers of the social behavior of the groups they seek to study are like eyewitnesses – list what people do and say they will do because "all behavior would be available to the researchers in life" with that of the historian whose "data restrict him to only partial aspects of cultural and social behavior, so he must reconstruct the other aspects from those he possesses."[35] It is not clear from this last passage whether or not Berkhofer thinks that those partial aspects are taken directly from the data, i.e., the evidence, without the application of reconstructive techniques, and if he does I should certainly wish to take issue with him. But I should want to say, in broad agreement, I think, with the spirit of his account in the chapter relevant to our present discussion, that what an historian can derive from the data which he confronts must surely be affected by what the theoretical social sciences have to tell us about the range of human and social possibility. With that said, we may go on to agree with Berkhofer – for this is what I take him to be saying – that historians might well have reason to believe more about the human past than may be constituted directly by thinking historically about their evidence. Should what can be directly constituted in the usual way turn out to be an action or event of the sort that social scientists have related theoretically to actions and events of other specific sorts, even though the evidence does not permit the direct constitution of instances of the latter sort, we may conclude that such did indeed take place. If we have theoretical reasons to believe that events of kind *A*, an instance of which may be constituted in the usual historical way by scholars working in some given field of history, are associated in some determinate way with events of kind *B*, for which we have no evidence or not evidence sufficient for an instance to be directly constituted historically, our account of the time and place in question might still make some general kind of reference – not an explicit description – of an event of the relevant kind if our confidence in the social-science theories involved was adequate to the occasion. Thus, it would appear that the social and psychological sciences have two roles to play in the task of historical constitution:[36] they define the range of human possibility within which the constituting activity of the historian is carried out, and they make possible the hypothetical postulation of possible events which might not otherwise be constituted given the limited character of the surviving historical evidence.

VII

Some may find the discussion of the present section rather unusual, yet since it continues our consideration of the way in which historical constitution is affected by intellectual commitments as to what is socially and humanly plausible, it is certainly not irrelevant to the main themes of the paper. Our point of departure will be some observations made by Carl E. Braaten on the subject of 'The Historical Event of the Resurrection.'[37] The question may obviously be raised as to what history might have to say about the resurrection of Jesus given how central the claim concerning it is to Christian belief. The presumption has been that Christians intend that the claim be afforded no less degree of historical factuality than is accorded other claimed events of antiquity, say Caesar's crossing of the Rubicon. Nor is it obviously clear how it could have the central significance it has for Christian belief and hope if its factuality were in any sense impugned or if it were treated as factual, if you like, but in a sense of factuality significantly unlike the factuality of things ordinary or mundane. Yet for a long time, there have been intellectual tendencies of precisely the sort, and increasingly we have witnessed the sharp separation between faith and factuality. No doubt one element entering into this divorce is the growing realization that the traditional proofs for the existence of God are of doubtful logical validity, a realization which tended to support among the faithful the view that their faith ought to be a freely given commitment coerced by neither fact nor logic. It is perhaps not fortuitous that one of the most trenchant critics of the proofs for the existence of God, David Hume, was also the author of an influential essay, 'On Miracles,'[38] which in effect argues that what is historically believable is what fits into the regularity of our experience. Yet essentially the same attitude was taken by a thinker so unlike Hume as F. H. Bradley, who says, "For everything that we say we think we have reasons, our realities are built up of explicit or hidden inferences,"[39] and "that an inference . . . is justified solely on the assumption of the essential uniformity of nature and the course of events."[40] Thus it is evident that a number of intellectual streams went into the forming of what Braaten calls "the world view of positivistic historicism" the control of which over historical method has resulted, in his view, in its being "understandable that theologians would become hostile or merely indifferent to it."[41]

Given the seemingly unchallengeable rupture of fact and faith,[42] the appearance on the scene of New Testament scholarship of existential hermeneutics,[43] initially inspired by the work of Rudolf Bultmann, who was in turn influenced by the philosophy of Heidegger, may well have seemed to Christian theologians as a promising way to make faith relevant to more than itself. Without entering into the details of it, which I am not qualified to do in any event, this new hermeneutical direction was less concerned – indeed, perhaps not concerned at all – with the factuality of the events of Easter than with the existential significance of faith in the Easter event, or, more particularly, what existential truth the assertions by the Gospel writers on the resurrection of Jesus contain. To take this direction is obviously not to be all that much concerned with the resurrection as an event like other events, and is to be impervious to the question of whether or not such an event is possible. But can one even talk about the existential significance of faith if the faith in question takes the form of asserting that something took place which could not have taken place? Can one actually avoid the question of the factuality of the resurrection and still seek to explicate its theological significance? That the answers to these questions must be in the negative seems precisely the point of Wolfhart Pannenberg's assertion that, "If one assumes that the dead cannot rise, that an event of this type can never happen, the result will be such a strong prejudice against the truth of the early Christian message of Jesus' resurrection, that the more precise quality of the particular testimonies[44] will not be taken into consideration in forming a general judgment."[45] Braaten goes on to add the following:[46]

> Pannenberg is right! Prejudices do have to be cleared away before the historian will possess the frame of mind to treat the historical evidences of the resurrection as "evidence." Perhaps historians have never acted more unprofessionally than when dealing with the New Testament testimonies to the resurrection of Jesus Christ. This happens when inflexible assumptions about the nature of man and of the world prejudge that such a thing could never have happened. The historical problem of the resurrection of Jesus thus has two sides, the historical testimonies themselves and the whole complex process which historians use in forming historical judgments about them.

If, for the most part, his theological colleagues are willing to leave the discipline of history to positivism or to naturalism, Braaten is not. What he prefers to see done is the reformulation of the criteria which govern what we take to be possible so that it becomes possible in principle to take seriously the factuality of Jesus' resurrection as an event like other events. Nothing less, in his view, is compatible with its having any significance at all for Christian belief. Thus, we see what Braaten's problem is, and we see it in such a way, namely, in the context of the broader problem of the logic of historical constitution, that the problem can make sense to us. Understanding the problem, however, does not make it easy to find a solution that Braaten could accept. For Braaten himself admits that he "cannot believe that a historian's judgment would lean weightily in favor of the historicity of the resurrection unless, *inter alia*, he were motivated to appreciate the historical basis of his actual faith knowledge of the risen, present Christ." And he goes on immediately to express his agreement with the following statement of Alan Richardson: "Apart from faith in the divine revelation through the biblical history, such as will enable us to declare with conviction that Christ is risen indeed, the judgment that the resurrection of Jesus is an historical event is unlikely to be made, since the rational motive for making it will be absent."[47] What Braaten wants is that theologians not yield up historical reality and research when talking about the resurrection and its significance. In his view, "Historical reality that is qualified by the resurrection of Jesus as its aim and meaning does not elude the methods of historical reason; yet reason needs faith as the dynamic of its vision." He goes on to add, "What reason sees is seen by reason, but *that* reason sees what it sees is made factually possible by faith."[48]

What shall we say to all this? Clearly, what Braaten wants is *not simply* that historiography be not so wedded to certain ways of limiting possibility as to preclude the possibility of human resurrection. And that is because he doesn't believe that historians will be open to that possibility – and thus be open to taking seriously and on their own terms texts which purport to contain the testimonies of witnesses who claim to have seen alive one whose death they had witnessed – unless they were already committed to the truth of still another claim that is logically independent of the first claim. And that is that one particular resurrection has certain extraordinary consequences beyond the mere return to life of some person who had died. In other words, while Braaten has formulated his problem in a way which easily fits into the framework of our own attention to the question of what the intellectual elements are which enter into historical constitution, the conclusion at which he arrives is that no one is likely to be open to accepting the historical factuality of the resurrection of Jesus who is not prepared to accept the Christian idea of its meaning. It is hard to doubt that he is right. But there is a point that is worth drawing from this conclusion. Historians may well agree on some matters of fact without agreeing on their significance. All may agree that on the established day and at the given place a certain action was carried out. Yet it is surely possible for there to be a number of opinions on what the effect of the action was, on how essential it was or was not to the events which followed it in time. I should think that this could be the case with respect to any event or action historians might constitute and concern themselves about. Braaten, however, says with respect to the claim of Jesus's resurrection that it is not an historical event which could be constituted by any historian who was not antecedently of the Christian opinion with respect to its significance. Thus, while he wants to rescue the concept of historical fact from the restrictions imposed upon it by positivist historicism so that the resurrection of Jesus can be accorded the same status *qua* historical fact as Caesar's crossing of the Rubicon, in the end he emerges with a resurrection which continues to be espistemically idiosyncratic.

VIII

I began this paper by stating that what I wanted to investigate was what it is that makes an historical reconstruction – or, as we have since learned to say, constitution – acceptable, and after all these thousands of words later, what can we say about that? One possible answer to the question in general, which I have not discussed earlier in this paper but which I have discussed elsewhere in order to defend,[49] has to do with what might be expected from a piece of historical constitution, what, so to speak, its logical function is in the context of historical inquiry. And that expectation is that it makes sense of the historical evidence. Given the character of the evidence we arrive at some conclusions with respect to what the historical past might have been like, and what in effect this means is that what the constituted historical event does is explain the evidence. And this seems to leave us with the problem of the logic of that sort of explanation. The logic of explanation tends to involve deduction, and if it is the function of the constituted event to explain the historical evidence, the presumption may be that the evidence must be deducible from the event as constituted. I have long thought that such deducibility is required by the logic of the problems, but I must confess that I have never understood how it could possibly be the case. Thus, to recur to an example discussed earlier, while I have never had any difficulty in appreciating the basis of Collingwood's conclusions with respect to the Irish colony in fifth-century Silchester, it was obvious from the start that given the assumption of such a colony no one would or could deduce therefrom the existence of a tombstone such as Collingwood described. We have already seen that the relationship between historical events and historical evidence is not deductive, and it would thus be exceedingly difficult to determine just how it might be argued that the logical function of the constituted event is to explain the evidence – though I must admit that I continue to believe that that is precisely what its logical function is.

For the present, however, we may drop that question entirely, because attention to the logical function of the event as constituted is not quite what we need here. Our problem, after all, is not with how the constituted event functions, but with how it is constituted, how the historian *comes to know* that such an event took place or arrives at the conclusion that *it is reasonable to believe* that some such thing happened. Far from talking about the relation between evidence and events already constituted, our concern throughout this paper has been with the antecedent stage during which the event emerges from or is constituted in the course of historical inquiry. And so far as that is concerned, I fear that what must be said is that there are no recipes for historical constitution. There are no rules or check-lists with reference to which we can know that the job of constituting some swath of the historical past has been done correctly. This much, however, was already apparent when it was recognized that historical constitution was not a determinate species of inference – along deductive or inductive lines – but was rather something like abduction. That kind of thinking is not the sort that applies settled canons of inference to agreed-to starting points in order to arrive at conclusions certified as sound by the correct application of the canons. It is, rather, a species of trial and error thinking, and if we are not able to discern in it a pattern of inference analogous to those of the better studied forms of thinking, we have, it seems to me, come to discover a number of its salient features. Its point of departure is the body of historical evidence. While we now see that the collecting and ordering of the evidence is not actually first in time inasmuch as what the evidence is which belongs together and is relevant to the solution of some problem of history is itself part of the attempt to deal with the problem – is itself affected by the direction in which the historian's work is taking him – yet with respect to some for-the-moment finished product of historical constitution its point of departure is a body of historical evidence. Obviously, if the work is successful, it has made sense of, explained the evidence – again the problem of the preceding paragraph. Yet it is clear from our discussion that we have to do with rather considerably more than the evidence itself, and the acceptability of an historical account to the community of scholars will depend not merely on whether or not it satisfies some conception of the logical relation which must exist between it and historical evidence. The plausibility to be sought in an historical account is not simply that of logical possibility. What can be believed is determined by the state of relevant knowledge in the historian's intellectual community, and this is no doubt one of the reasons which precludes any effort to see the problem as *directly* one of deriving the historical past from the historical evidence, for the derivation is *mediated* by the intellectual and theoretical conditions of plausibility. The above discussion of the relevance of social and psychological sciences to historical constitution were intended to bring this out, and so was that of Braaten's claims for the resurrection of Jesus as properly within the domain of a history purged of its naturalistic inclinations. The two discussions in fact complement one another in the following way. The account of the role of the social and psychological sciences points to how our conceptions of what is believable about human affairs come to develop, and how sophisticated common sense is affected by the outcome of systematic research in the relevant fields. If the conditions of plausibility are independent of the historical evidence and represent the introduction of elements seemingly extraneous to the historical past, at least we have the increasing assurance that those elements need not be relative to the subjective idiosyncracies of individual historians, but may, rather, be rooted in the best available knowledge of the time. Our discussion of Braaten's problem, however, led to the conclusion that the resurrection with which he emerges is epistemically idiosyncratic. In the context of this discussion, what that means is that without the intrusive factor of faith, Braaten cannot achieve the results he requires to have. And that, of course, undermines the claim he is making on behalf of the resurrection to begin with. What he had not managed to do was present the sort of argument that would have enabled historians to take seriously that a resurrection – any resurrection – is a plausible historical possibility and that the texts which contain what purport to be testimonies concerning the resurrection of Jesus *could be* sound evidence for what they claim. Thus Braaten does not really succeed in enlarging the realm of acceptable possibility for historians.[50]

The historian, then, must work within the limits imposed by the evidence and by prevailing conceptions of historical plausibility. Within those limits, he must persuade his colleagues that the evidence is to be ordered in the way he proposes as relevant to the problem he seeks to deal with, and that, in consequence, what he presents in his book or essay is precisely what it is reasonable to believe took place once upon a time.

State University of New York at Binghamton

NOTES

* [This paper was received and accepted for publication February 2, 1976 – Ed.]

1 Ch. V, Langlois and C. Seignobos, *Introduction to the Study of History* (London, 1898).

2 This is made patently clear by recent work on sixteenth-century French historiography; cf. Donald R. Kelly, *Foundations of Modern Historical Scholarship* (New York and London, 1970); and George Huppert, *The Idea of Perfect History* (Urbana, Chicago, London, 1970).

3 For a suggestive and stimulating conception of historical synthesis remarkedly unlike that of Langlois and Seignobos with a somewhat Kantian orientation, see Henri-Irénée Marrou, *The Meaning of History* (Baltimore and Dublin, 1966).

4 A handy guide to the development of Husserl's interest and work on this matter, as well as references to his relevant writings, is Robert Sokolowski, *The Development of Husserl's Concept of Constitution* (The Hague, 1964).

5 See Alfred Schutz and Thomas Luckmann, *The Structures of the Life-World* (Evanston, 1973).

6 To avoid unnecessary digressions, I am willing for the moment to say that we know nothing of that part of the human past which lies beyond memory other than what has emerged from historical research. If it were pertinent to raise the matter here, I would point out, however, that our memories have no presumption of being veridical, and that historians of the most recent past, even if they were participants in the events they deal with, may not simply transcribe their recollections into their texts, but must treat them critically as if they were documents. I have some more extended observations on

this subject early in the chapter called 'The Narrativist Thesis' of my book *Historical knowing* (Austin, Texas, 1976).

[7] I have tried to say why in the chapter entitled 'Historical Objectivity' in the book referred to in the previous note.

[8] Allen Johnson, *The Historian and Historical Evidence* (New York, 1926), contains a useful and informative survey of such things.

[9] Collingwood's contribution to R. G. Collingwood and J. N. L. Myers, *Roman Britain and the English Settlements*, 2nd ed. (Oxford, 1937).

[10] In calling it an "historical fact," I mean only that it is the outcome of historical research or an act of historical constitution. This does not commit me to belief in its truth or its being reasonable to believe in any other way. (Oddly, this usage is to be found in Langlois and Seignobos, p. 205, where they speak of "discordance between facts" established according to different methods.) This matter is discussed in greater detail in the chapter called 'Historical Facts' in the book referred to in note 6.

[11] Collingwood and Myers, p. 316.

[12] Thus, in discussing some views of the origins of certain Jewish festival lights, T. H. Gaster (*Commentary*, December 1952, p. 536) says: "Whether any of these theories is right cannot now be known; all of them are based on deductions from fragmentary and inconclusive evidence."

[13] The discussion which follows is based upon Hjalmar R. Holand, *Norse Discoveries and Explorations in America, 982–1362*, (New York, 1969, Dover Publications ed.), part II.

[14] Actually, the word for lake is not contained in the text, but given what skerries are and that we are dealing with an area in the midst of the continent and not on its coast, it is obvious that a critical analysis of the text requires that it be read so as to refer to a lake.

[15] Including the claim that the Mandan Indians have Scandinavian features.

[16] Should it turn out that other scholars were working in the way that Holand was, and that there was already discussion of the possibility that remnants of the Paul Knutson expedition had ended up in the region of the Kensington Stone, it would not be difficult to amend my statement so as to take account of it. The personal contribution of Hjalmar R. Holand is of no particular relevance to the matter under discussion.

[17] The usefulness of Holand's book as an example of the sort of intellectual work we are interested in in no way depends on its being acceptable to the community of scholars in the long run. That the outcome of his constituting activity is acceptable obviously depends, among other things, on the authenticity of the Kensington inscription, and it appears that the overwhelming majority of philological and runological opinion rejects it. If this opinion is sound, then it must prove to be the case that the inscription is a modern hoax. An interesting account of the evidence for that may be found in Theodore C. Blegen, *The Kensington Rune Stone* (St. Paul, Minn., 1968), a contribution, as it were, to the constitution of an event rather different from what we find in Holand's book. And one which would have served our purpose as readily as his, though one which would have been rather more difficult to present in brief compass. But on the assumption that the inscription is what it purports to be, Holand's use of it and other data provides a perfectly serviceable example of the sort of constituting endeavor which concerns us in this paper.

[18] For a somewhat opposing view, see Maurice Mandelbaum, *The Problem of Historical Knowledge* (New York, 1938), p. 252. In its context Mandelbaum's account seems reasonable, but that is only because he has chosen to discuss an example from very recent history where the classification and ordering of evidence is much less of a problem than in other periods.

[19] To talk of two sets, even though one consists only of the Kensington Stone, is not unreasonable. Even though Holand begins with the stone, by the time he is finished he makes reference to other objects found in Minnesota which he takes to be Scandinavian in origin. It could easily have happened that some scholar took note of the existence of a collection of artifacts found in America which seemed more to resemble objects undoubtedly of medieval Scandinavian provenance than anything else known from North America. In that event, when Holand got started he would have been able to begin, not with one seemingly-out-of-place inscription, but a whole body of seemingly-out-of-place artifacts. His problem, in the end, would be the same, namely, to constitute a swath of the historical past that made sense of the seemingly-out-of-place.

[20] A clear and concise survey of the area may be found in Wesley C. Salmon, *Logic*, 2nd ed. (Englewood Cliffs, 1973), ch. 3.

[21] S. F. Barker, *Induction and Hypothesis* (Ithaca, 1957), pp. 3 and 10.

[22] C. Behan McCullagh (*History and Theory* **12** (1973), p. 453) says that "the normal form of historical inference" is the statistical syllogism, but I find this claim completely mystifying and have no idea as to how it could be made out. I suppose some clue to what he has in mind is to be found in his earlier discussion, p. 441; but that discussion seems to involve statistical regularities between classes of historical evidence and kinds of historical events the like of which has never figured in the attempts of historians to constitute the human past.

[23] About induction, Peirce says such things as the following: "Induction consists in starting from a theory, deducing from it predictions of phenomena, and observing those phenomena in order to see *how nearly* they agree with the theory" (5.170; italics in original). Elsewhere, he says, "The third way is *induction*, or experimental research. Its procedure is this. Abduction having suggested a theory, we employ *de*duction to deduce from that ideal theory a promiscuous variety of consequences to the effect that if we perform certain acts, we shall find ourselves confronted with certain experiences" (8.209; italics in original). On previous occasions I have argued that the logical role of an "historical reconstruction" is to explain the historical evidence in the sense that the "reconstruction" is an hypothesis from which the evidence ought to be deducible (cf. 'A Note on the Status of Historical Reconstructions,' *Journal of Philosophy* **55** (1958), 473–79; and 'Evidence and Events in History,' *Philosophy of Science* **29** (1962), 175–94, parts I and II). I still subscribe to this view. Thus it would seem that I subscribe to a view with respect to the relationship between evidence and events in history which Peirce would surely call inductive. The constituted event functions as an hypothesis, and from it one must not only be able to deduce the evidence in hand which suggested it, but derive some conception – I hesitate to use "deduce" here – of what additional supportive evidence, or kinds of supportive evidence might be hoped for. But this relation between evidence and event can only be discovered after the event has been constituted. Our problem here concerns the nature of that constitution. And that is precisely what Peirce himself says in the second of the two passages just quoted when he indicates that a theory is first suggested by abduction and then the inductive task of confirmation is undertaken.

[24] For the sake of those who would want to know, I may point out that volume VII of Peirce's *Collected Papers* contains a chapter called 'The Logic of Drawing History from Ancient Documents' and a section of that chapter (7.136–143) is concerned with abduction. In spite of having a title which suggests its relevance to what I am attempting here, I did not find Peirce's discussion of particular use. There are two reasons for this. His conception of drawing history from ancient documents deals with ground for believing or rejecting ancient testimonies, and it betrays no awareness of the problem of how the historian constitutes a past he does not copy out of surviving texts. And he thinks that the only way to avoid idiosyncratic subjectivism in determining what it is reasonable to believe is to be able to apply the calculus of probability, though his own practice (cf. 7.235–236) is not always consistent with this belief.

[25] Norwood Russell Hanson, *Patterns of Discovery* (Cambridge, Eng., 1958).

[26] A. M. MacIver, 'The Character of a Historical Explanation,' *Aristotelian Society Supplementary Volume* 21 (1947), p. 38.

[27] Marrou, *op. cit.*, p. 314.

[28] Jean Schneider, *La ville de Metz aux XIIIe et XIVe siècles* (Nancy, 1950).

[29] Marrou, *op. cit.*, pp. 314f.

[30] The concrete or elementary facts that Marrou mentions are clearly intended to be the facts contained in the purified declarative sentences which emerge from the Langlois and Seignobos methodology.

[31] I attempt to show why in the final chapter of *Historical Knowing* (see note 6), but there is no need to present that argument here.

[32] Which, of course, he could not, for want of suitable historical techniques.

[33] John W. Hall, 'The Muromachi Age in Japanese History: A Report on the Conference Held in Kyoto, August 27–September 1, 1973,' *Social Science Research Council Items* 27 (December 1973), 41–46; esp. p. 41.

[34] Robert F. Berkhofer, Jr., *A Behavioral Approach to Historical Analysis* (New York and London, 1969), ch. 5.

[35] *Ibid.*, p. 99.

[36] Of course, they must figure centrally in attempts to *explain* historical facts, but that subject is not our concern here.

[37] Carl E. Braaten, *History and Hermeneutics* (New Directions in Theology Today, vol. II) (Philadelphia, 1966), ch. 4.

[38] Which is chapter 10 of his *Enquiry Concerning Human Understanding*.

[39] F. H. Bradley, *The Presuppositions of Critical History* (Oxford, 1874), p. 10.

[40] *Ibid.*, p. 15.

[41] Braaten, *op. cit.*, p. 43.

[42] Except, presumably, among fundamentalists in religion who are not or not much affected by secular intellectual currents.

[43] Braaten, *op. cit.*, pp. 37–42; but see, too, James M. Robinson's lengthy, but detailed and informative, 'Hermeneutic Since Barth,' James M. Robinson and John B. Cobb, Jr. (eds.), *The New Hermeneutic* (New Frontiers in Theology, vol. II), (New York, Evanston and London, 1964).

[44] As reported in the Gospel accounts: Pannenberg's concern is with "the presuppositions under which the historical question could be taken seriously" (Braaten, *op. cit.*, p. 97).

[45] Quoted in *ibid.*, p. 98.

[46] *Ibid.*

[47] *Ibid.*, p. 102.

[48] *Ibid.*; his italics.

[49] See my papers cited in n. 23 and 'Collingwood on the Constitution of the Historical Past,' in M. Krausz (ed.), *Critical Essays on the Philosophy of R. G. Collingwood*, (Oxford, 1972), pp. 262ff.

[50] I remember attending a session at one of the annual meetings of the American Philosophical Association, Eastern Disvision – I no longer remember the year – in which the consequences for philosophy of recent research in extra-sensory perception was being discussed by C. J. Ducasse and J. B. Rhine. The latter tried to insist that what was called for was a radical break in the scientific world-view along some sort of spiritualist lines. The former thought that it would be possible to accommodate the findings of the research in question by means of a modification of the prevailing general orientation in the direction of some sort of subtle scientific materialism. I no longer recall the details of either position, but the relevant point is that only if something along the lines of Ducasse's proposal could be worked out could the scientific respectability that still seems to elude Rhine's field be won in the present intellectual environment. Braaten's problem is similar.

CAROL C. GOULD

BEYOND CAUSALITY IN THE SOCIAL SCIENCES: RECIPROCITY AS A MODEL OF NON-EXPLOITATIVE SOCIAL RELATIONS*

The problem to which I address myself in this paper is the construction of an adequate model of explanation in the social sciences. Causal models of explanation have been criticized as inadequate in their conception of human action. Such criticisms have been offered by, among others, Anscombe, Peters, Melden and Hamlyn in their critique of behaviorism, Winch in his critique of causal explanation in sociology, von Wright in his discussion of explanation in history and the social sciences, and of course Husserl, Heidegger and Sartre in their analyses of consciousness, *Dasein* and choice. In general, I am in agreement with the well-known criticisms which these philosophers have offered, but I want to go beyond them in various ways. In particular, the criticism of causal models which I shall propose in the first part of my paper will focus not only on the inadequacy of these models in their account of human action, but also on their inadequacy as accounts of interaction. However, in rejecting causal explanations of human action and interaction, I do not intend to exclude causality from the social sciences. Rather, I will claim that human agency is causal insofar as it acts on non-human objects. As such, causality will be seen to be a delimited but important aspect of a total explanation in the social sciences.

To put this somewhat differently, my claim is that while the category of causality is applicable to the relation of subject to object, it is inapplicable to the relation of subject to subject. For the understanding of this relation of interaction, which is the primary subject matter of the social sciences, an entirely different category is necessary. I will propose that this category is that of reciprocity.

The concept of reciprocity has been curiously absent from almost all discussions in the philosophy of the social sciences. However, it has been present in curious ways and for a long time in the social sciences themselves and especially in sociology and cultural anthropology. Thus, for example, it is an important concept in the work of Mauss[1], Malinowski[2], Mead[3], Simmel[4], Schutz[5], Piaget[6], Parsons[7], Homans[8], Blau[9], Gouldner[10], and Goffman[11], among others. Although the concept of reciprocity has entered into the work of these theorists in many different ways, it has primarily been interpreted in terms of systems of reciprocal exchange, whether economic,

R. S. Cohen and M. W. Wartofsky (eds.), Epistemology, Methodology and the Social Sciences, 53–88.

social or psychological. In these accounts, reciprocity is seen to function in acts of exchange, in which either the acts themselves or the objects exchanged, are taken to be equivalent. However, in reviewing the social science literature, it becomes evident that the concept of reciprocity is not clear and that its use presents many problems for conceptual and methodological analysis. My aim in the second part of this paper will be to clarify the concept by developing a systematic philosophical model of reciprocity. I shall work out this model in the first place through an analysis of antecedent formulations of this concept in the philosophies of Kant and Hegel and in the social theories of Hegel, Marx, Simmel, Schutz, Gouldner and Habermas and then proceed to the systematic formulation of my view. I shall propose a concept of reciprocity in which it serves both as a model for explanation and as a normative and critical principle, namely one which poses reciprocal recognition as a condition for the full development of freedom. It is my claim that such a concept of reciprocity is an essential part of an adequate explanation in the social sciences and that it replaces causal models of explanation in the domain of social interaction.

Both in my critique of causal models and in my account of the model of reciprocity, I base my approach on an analysis of the nature of the basic entities which make up society and of the nature of their relations. Such an analysis is a fundamental aspect of what I would call a social ontology. I take these basic entitites to be agents who stand to each other in social relations. On this view, these agents are understood as social individuals who cannot be taken apart from the relations in which they stand to each other. On the other hand, they are understood as concretely existing and acting individuals, who cannot be reduced to their social relations and roles.

PART I

I shall begin with a critique of causal explanations in the social sciences and propose an alternative conception of causality as it concerns human action. Since I intend this first part of the paper to serve only as background for the analysis of reciprocity, this discussion of causality will be brief and schematic. I should state at the outset that my discussion will be restricted to causality as it concerns human affairs, and will not deal with the question of causality in nature. I am not proposing any theses about the use of causal explanations in the natural sciences nor about the relation between causality in human action and causality in nature.

The motivation for a critique of causality stems from the serious problem that causal explanation poses for an account of human agency. In particular, in its various versions, causal explanation of actions may be seen to be incompatible with the freedom of agents. Such causal explanations in the social sciences, and the methodological discussion of them, take a variety of forms. We may discern three main types of such causal explanation. First, there is the view that human action, whether individual or social, is determined by forces or circumstances external to it, where these external causes are variously understood as natural or social. Such a view interprets causality as productive, that is, as actively producing or bringing about actions, whether directly or through determining the natures of the agents. This view may be associated with both the nature and nurture alternatives in social and psychological theory, in which either external forces or circumstances, on the one hand, or innate biological or psychological structures, on the other, are held to determine behavior. One might mention here, for example, such biological determinists as Konrad Lorenz and E. O. Wilson; such economic determinists as Charles Beard and those who advocate vulgar or reductive forms of Marxism; and such psychological determinists as Freud in some of his writings. One should of course add to this list those theorists who hold that race or sex determine one's nature and the character of one's actions.

A second and more common conception of causal explanation is that which identifies the causal relation with invariant correlation or law. Here the determination of human actions is interpreted in terms of the relations between dependent and independent variables, where this relation is seen as law-governed. Even where the laws concerning human actions are understood as statistical regularities, this view interprets this as a limitation imposed by the complexity of the variables or the imperfect state of our knowledge. In this way, this view maintains a deterministic approach to human actions. Moreover, despite its agnosticism with respect to necessary connection, this view emphasizes that knowledge of the laws of human behavior permits the prediction and control of such behavior. Among those social scientists who aim at a causal account of action in this second sense one may mention the Skinnerians in psychology, Blalock, Homans and also Weber (in some of his methodological writings) in sociology; and more generally, the positivists and logical empiricists who follow the nomological model of explanation in the social sciences and in the philosophy of the social sciences.

The third variant on the causal model of explanation is the holist view that social structures, norms, or roles determine human actions. Even where the understanding of the agents is taken to be required for an explanation of their actions, this understanding itself is seen as bound to or determined by social structures or norms imposed on the agents from without. Some examples of this view are to be found both among structuralists and functionalists, such as Durkheim, Parsons, Lévi-Strauss and Althusser. This view seems to me to fall within the causal model inasmuch as it conceives the forms which human action can take as determined or caused by the structures which are external to agents or their actions.

Thus there are three main types of causal explanation in the social sciences, namely, productive causality, functional correlation interpreted as law-governed or nomological, and holism. I would hold that despite their differences, all three of these views may be criticized for their failure to recognize the distinctive character of human action as free and intentional. The criticisms which may be made of such causal models may be summarized briefly here.

The first set of criticisms are aimed at showing the incompatibility of causal or deterministic explanations with the very concept of an agent, or of an action. These well-known criticisms have been developed by the post-Wittgensteinian action theorists, such as Anscombe, Melden, Peters, Hamlyn and Kenny, as well as by phenomenologists and existentialists, such as Husserl, Heidegger, Sartre and Merleau-Ponty. One argument along these lines is that it is contradictory to assert, on the one hand, that an action is something done by an agent and on the other hand, to assert that his or her doing it is determined by an external cause. The concept of doing something or of an action entails choice or that the agent initiates the action. Furthermore, the concept of choice entails that an agent could do otherwise. By contrast, causal explanation in any of the three major senses described above – namely, as productive, nomological or holistic-structural – requires that an action be understood as the effect of antecedent causes, such that the agent does not initiate the action and could not do otherwise, given the initial conditions. The experience and intuition of choice are thereby reduced to mere appearance or subjective error. Such causal accounts thus fail to do justice to the distinctive character of human action as free.

As the critics of causal explanation have also pointed out, causal accounts fail to grasp yet another feature of action, namely that it is intentional. That is, action is directed towards the realization of the conscious purposes of an agent, although these purposes need not be reflected upon or deliberated on in all cases. Intentions or purposes are what characterize actions as not merely bodily motions and also constitute and identify an action as being the particular action that it is. The individuation of actions therefore essentially requires the distinctions among intentions which are excluded from causal accounts. Thus, such explanations of human action which fail to take into account its intentional character cannot be adequate to the very nature of the subject matter they seek to explain. But causal explanations do violence to the subject matter in just this way.

An additional criticism of the causal view along similar lines is that the very act of offering a causal explanation of actions presupposes intentionality and choice on the part of the one who offers it. It would therefore be self-refuting for a philosopher or scientist to argue that actions are to be explained causally when the very act of giving such an argument is itself an instance of conscious purpose which involves choice. Forms of this argument have been given by Husserl, Jaspers, and more recently by Toulmin. Thus, for example, in his essay 'Philosophy as Rigorous Science,' Husserl points out that the premises of causal explanation are refuted in the practice of the scientist or the philosopher "whether it be in constructing theories or in justifying or recommending values or practical norms."[12] According to Husserl, this very activity of explaining, theorizing and justifying presupposes intentionality. Thus it is implied in his view that causal theories of action are self-refuting.

Another criticism of causal explanations related to the arguments from intentionality offered above is that human actions and social phenomena in general cannot be understood or explained apart from the meanings which they have for agents. Such an argument has been made by Winch and the hermeneuticists among others. Inasmuch as causal explanations take the relations among the entities which they study to be external, they cannot take into account the understanding which agents have of the meaning of their own actions or of the actions of others. In this way, they again fail to be adequate to the subject matter of their study.

Yet, there are certain theorists who do take the meanings of actions as central, but who nevertheless adopt a causal interpretation of these actions. In particular, they take the meanings themselves as determining the actions, where these meanings are taken to be rules or norms of action. This view, associated with certain holist theories, takes agency to be constrained or directed by prevailing structures or norms of permissible action. This particular version of causal explanation fails to recognize that agents not only have the choice of acting or forbearing from acting, in accordance with a pre-determined norm or rule, but also can choose to alter these norms themselves and to discover alternative ways of acting. And indeed, history reveals just such a successive replacement of norms and rules.

I would claim that it is indeed a basic feature of action that choice is not limited to a given set of alternatives, but rather that agents can discover or create new alternatives. This feature of action is one of the bases for historical and social change, as well as for individual self-transcendence, both of which are fundamentally related to the nature of freedom. I am not claiming that agents always do in fact discover or create new alternatives, or that they always do in fact transcend themselves or realize their possibilities, but rather that they always have the capacity to do so. In this way, freedom as such a capacity is presupposed in all action, whether individual or social. In this respect, I am in agreement with the phenomenological view, especially as it is expressed by Heidegger and Sartre, in its emphasis on the character of human activity as fundamentally free.[13]

The arguments which I have thus far presented show the inappropriateness of causal models in explaining the actions of agents. But these very same arguments also have an important consequence for the account of interaction among agents, a consequence not developed by the action theorists and the phenomenologists. If it is true, as I have argued, that the actions of agents are not caused at all, then they are also not caused by the actions of other agents. The interaction of agents is thus not a causal relation and cannot be explained causally. It follows that interaction requires a different analysis and a different mode of explanation. I shall propose in the second part of this paper that the concept of reciprocity provides the key to such an alternative account. However, it should be mentioned here, that to deny that the actions of agents can be caused by other agents is not to say that agents may not be coerced or constrained by the actions of other agents. In fact, such coercion or constraint is a significant feature of social interaction and will also be discussed in the second part of this paper.

Furthermore, to deny that agents cause the actions of other agents is not to say that agents are never causal in their actions. Indeed, I would propose that a conception of causality is needed for the domain of human action, but it is one which differs from the conceptions given in the various causal models sketched above. My account also differs from that given by action theorists who, while denying that actions are caused, yet fail to deal with the respects in which actions are themselves causes. This alternative view of causality is derived from my reconstruction of Marx's account of human activity as the productive and transformative activity of agents.[14] My account of the causality of agency may be sketched briefly here.

On this view, agents are causally efficacious in their activity of transforming objects in accordance with their purposes. Such activity may be generally characterized as productive, in that it involves making or creating new objects. This activity is one of shaping or forming material for the sake of realizing some human purpose or end.[15] One may characterize this causal relation as a subject-object relation, as distinct from a subject-subject relation; that is, in this case, the objects are things and not other persons. The subject or agent in such a causal relation may be either an individual or a group of individuals. Where there is such a group of individuals, they may be said to constitute a common cause. Thus, when I use the term agent, it should be taken to connote either a single individual or a group of individuals acting together.

This causal relation may be analyzed as an asymmetrical internal relation. It is an internal relation, in the sense that both the agent and the object are changed in the relation. That is, through this productive activity, a given material is changed by being worked on; it becomes a different thing from what it was before. Moreover, the nature of the object produced is internally related to the nature of the activity involved in its production. For example, in making a chair, the particular form of the chair will depend on the chair's intended use and its specific character will depend on the instruments employed in its production, such as plane, chisel or saw, as well as on the specific materials such as wood, plastic or metal, which the agent chooses to use. Similarly, the nature of the agent's activity will depend not only on his or her purposes, but also in part on the materials and instruments used. Thus not only the object but also the activity is changed in the relation. Moreover, insofar as the mode of activity becomes a characteristic one for the agent – for example, when a certain skill is acquired – the agent also is changed by the activity. Therefore, causality is not to be understood as an external relation, but rather as an internal one.

However, my account preserves the asymmetry of the causal relation. That is to say, although both the subject and the object are changed in the relation, only the subject causes changes in the object, whereas the object causes no changes in the subject. Rather, objects provide what I would call the objective conditions for the agent's activity and for the agent's self-change. Such objective conditions are thus not to be construed as causes of action. Yet such conditions of action have often been construed as themselves the causes of action. That is, actions have been taken to be causally explained as the consequences of the set of antecedent necessary and sufficient conditions. I would argue that the objective conditions for an action are never more than the necessary conditions for that action and are never jointly sufficient to produce that action. Objective conditions are therefore not to be construed as causes of action. In fact, objects come to be conditions only with respect to given purposes and to given modes of activity of agents. Thus for example, the availability of a certain material, such as wood, and of certain instruments, such as saws or chisels, is not itself the cause of the activity of making chairs. Rather, they are to be regarded as objective conditions, in the sense that the chair could not be made without them or something else which could be substituted for them. Such objects may therefore be regarded as enabling conditions for human activity.

But the question arises: What of constraining or disabling conditions? Don't these have a causal effect on actions? Here, too, it seems to me that one must distinguish such conditions from causes. The absence of those conditions which may be necessary for a given action, or which constitute the means for its fulfillment, may make a certain course of action impossible, at least until these conditions are provided or an alternative to them is found. However, the absence of these conditions does not itself determine or cause a given course of action. Under these circumstances, an agent may choose to modify or abandon an envisioned course of action. On the other hand, an agent may decide on a course of action which is intended to change the conditions or to overcome the constraints. In this case, neither the original constraining conditions nor the newly created enabling conditions are causes of the agent's action. Rather, in such a case, the agent has overcome the constraints and has created his or her own enabling conditions through his or her own causal activity. That is, in this process of overcoming constraints, the agent is concretely self-transcending or free.

There is a sense in which not only objects but also other persons may be said to be conditions for the actions of an agent. Thus an agent may require the activity or forbearance of other agents in order to realize his or her purposes. For example, one person often requires the skills, ideas and capacities of others or their support in order to carry out a course of action. Or again, the permission or forbearance of others may be required as a condition for acting. However, I would propose that the relation of an agent to other persons in these respects should be distinguished from the relation of an agent to the material conditions of his or her activity. As I will argue later, the relation to others as social conditions for one's activity requires an analysis in terms of reciprocal or non-reciprocal interaction. Yet, this is not to say that the material conditions for agency are not themselves social. For most often, the material conditions are the products of the past activity of other agents.

Moreover, among the conditions for agency are not only material conditions but also what we may call objective social conditions. These include institutions, organizations, social rules and practices, laws, etc. These social conditions are also the creations of agents. They are the products of past activity and as such constitute part of the historical and social conditions of the present activity of agents. They are the arrangements through which agents carry out their activities and interact with each other. However, here too, these social conditions are not to be taken as causes, but rather as enabling or constraining conditions for human agency, which, as with material conditions, can be changed by the agents.

However, although conditions are not causes of actions, they are nevertheless very important aspects of a complete explanation of human actions.[16] In order to understand an action, one must know the conditions, both enabling and constraining, in which the action takes place. An account of the initial conditions is important both for understanding the presently available resources for action and for understanding the choices which an agent may make to change these initial conditions when these conditions are not adequate for the realization of an agent's purposes. Further, since agency is causal, a full account requires a consideration of the ways in which it acts on conditions. Indeed, I would make a proposal for the methodology of the social sciences along these lines, namely that in addition to an account of the intentionality of human actions, a full explanation in the social sciences should also include among its basic features an account of the ways in which human agency is causal and an account of the conditions under which such causal agency takes place. However, these three features alone are not sufficient for an explanation in the social sciences. Such an explanation also requires the inclusion of a fourth basic feature, namely, the actions of agents with respect to each other. This is the feature which I call reciprocity and which I will discuss in the second part of this paper.

In the light of the foregoing discussion, one may analyze the error which is made in causal models of explanation in the social sciences in the following terms: First, where objective conditions are regarded as causally effecting the actions of agents, one may say that the mistake is to attribute to objects or institutions as products of human activity the agency which only subjects have. Second, where the causal models attribute causal efficacy to the relation of an agent to another agent, that is, of subject to subject, the error consists in taking the subject acted on as if he or she were an object.

I would propose that both of these misinterpretations of causality are not merely philosophical or methodological errors. Rather, I would argue that they have an ideological dimension, in that such interpretations uncritically incorporate at the level of theory a particular form of social relations characteristic of a given historical period. That is to say, such causal views in part express or reflect in theory the forms of domination which characterize economic and political life in capitalist societies. In these contexts, relations among agents appear in reified form as relations among things. The causal efficacy of agents is attributed to external objects or institutions which are held to operate according to objective causal laws, independent of the agents. The agents themselves are regarded as subordinated to these laws. Further, this leads to the view that their actions are subject to prediction and control. The agents are thus treated as though they were objects, standing in external relations to other objects. The actions of agents are thus seen as caused by the very products of their own activity. Furthermore, the relations among agents are understood in terms of the relations among their products. This characteristic of social relations under capitalism is what Marx described as the fetishism of commodities. Thus in the *Grundrisse*, he writes, "... Their own exchange and their own production confront individuals as an objective relation which is independent of them."[17] Further, he says that the activity of individuals in production appears to them "as something alien and objective, confronting the individuals, not as their relation to one another, but as their subordination to relations which subsist independently of them."[18]

The attribution of causal efficacy to institutions and material conditions may be seen to derive from the real power which such factors have in social life. Similarly, the attribution of causal efficacy to agents over the actions of other agents derives from the real facts of social domination and exploitation. But it is a mistake to interpret such relations of power and domination as causal relations. Rather, what I would propose (and will discuss further in the second part of this paper) is that the social power over the actions of agents which is exercised by other agents and through the medium of institutions consists of their control over the conditions for the agency of others. What I mean by this is that in the leading forms of domination, exploitation or coercion in social life, one individual or group of individuals is in control over or has disposition over the conditions which another individual or group of individuals requires as means for the fulfillment of their purposes. But since conditions are not causes, as we have seen, to control the conditions for action is not to cause the action.

Insofar as causal models of explanation in the social sciences interpret social relations or the workings of social structures and institutions on the model of causal laws, they tend to see such historical social relations as if they were in the nature of things. In this way, they rule out the possibility of changing such social relations and institutions by means of free human action. To the extent that they do this, such causal theories may be said to contribute to the ideological legitimation of existing relations of domination and exploitation.

PART II

Having criticized causal models of explanation in the social sciences in the preceding part of my paper, I should now like to turn, in this second part, to the question of what lies beyond causality. Namely, what model of explanation should replace the causal account of the relation between subject and subject? I would like to propose and develop the concept of reciprocity as the appropriate model for the explanation of social interaction.

In what sense would the relation of reciprocity serve as a model of explanation in the social sciences? The received view of explanation in the sciences is the nomological or hypothetico-deductive model in which singular statements describing events or phenomena are said to be explained when they are deducible from universal or law-like statements together with statements of antecedent conditions. But as I have argued earlier, this would require that human actions be law-governed and this entails a causal account of action of the sort I have rejected. The alternative phenomenological account of human action and also some forms of action theory explain such actions on the basis of an understanding of the intentions of the agents. Although I agree with this requirement for the explanation of actions, as I have indicated, it is inadequate in that it limits the explanation of action to the intentions of individual agents. Thus on my view, it fails to give an adequate account of the social dimensions of action. Finally, where the explanation is sought in the rule-following or rule-governed behavior of individuals and hence implies a social dimension, the explanation concerns the relation between an individual's understanding and the rule rather than the relation of interaction among agents. By contrast, the model of explanation in terms of reciprocity which I propose here takes as its fundamental unit a social interaction. Thus, it focuses on a crucial feature of action which is neglected by the other models of explanation. However, I am not claiming that reciprocity exhausts the required features of an explanatory model. Rather, it forms part of a more complete model of explanation which would also include an account of the intentionality of agents, the causality of human agency and the objective conditions of action as I described these earlier.

The conception of reciprocity which I will set out here is founded on the conception of agency or human action presented above. Furthermore, just as causality, understood as agency, was seen to be compatible with freedom as a human capacity, so here too, reciprocity is understood as compatible with the freedom of agents. Indeed, my conception of reciprocity is intended to serve not only as a principle of explanation, but also as a norm. That is, reciprocity in its most developed form will be seen to be a condition for the full realization of freedom. In fact, however, the study of social interaction reveals forms of social relations which are characterized by domination and exploitation, in which the relations are non-reciprocal. Thus, my account will stress both reciprocity and non-reciprocity in social life, where reciprocity is understood as the norm in terms of which a critique of non-reciprocal social relations may be developed. Furthermore, the conception of reciprocal and non-reciprocal social interaction to be developed here sees them as historically changing, that is, as taking on different forms in different stages of historical development.

It should be noted that although I am proposing a model of reciprocity as central for understanding social interaction, it does not exhaust all the features of such interaction. Specifically, the cases of joint action or shared experiencing, though they presuppose reciprocity, yet involve other considerations which I will not discuss in this paper.

In what follows, I will develop my conception of reciprocity through a consideration of the antecedents of this concept. What I am doing here is to construct a model. Such a construction proceeds by selecting out the theoretical entities and relations which are taken to be most relevant to the explanation of a given domain and by developing a coherent or systematic structure which gives the most plausible account of the phenomena to be explained. In my construction, I will pick out those elements in previous accounts of the concept of reciprocity which serve the purposes of understanding social interaction. I will do this in the form of an analytical exposition rather than a critique of some earlier views. I will bracket the detailed arguments for or against the elements which I choose. First, I will briefly consider the development of reciprocity as a formal concept in philosophy in the works of Kant and Hegel. It may be mentioned that Aristotle has a striking discussion of reciprocity in his *Nicomachean Ethics*, both in connection with justice and equivalent exchange in Book V, and in connection with friendship and love in Book VIII. But for the sake of brevity,

I will not deal with his discussion here. After considering Kant's and Hegel's views on reciprocity, I shall sketch some of the ways in which the concept of reciprocity has been developed in social theory, as a category of social explanation. Of interest here are the social theories of Hegel, Marx, Simmel, and Schutz and more recently, Gouldner and Habermas. In considering these antecedents, I do not intend to give a comprehensive or a historical account. Rather, I plan to use them only as a foil for the development of my own view. I shall then consider briefly some of the ways in which the concept of reciprocity is presently used in the social sciences and make a few methodological suggestions about how it should be used.

I shall begin with a consideration of the philosophical analysis of the concept of reciprocity which is given by Kant. In the *Critique of Pure Reason*, Kant introduced reciprocity as the third category of relation (where the first two are substance and causality). He defines it as the ground for the possibility of our experience of the coexistence of independent entities as parts of a whole or a totality. By contrast to the category of causality, Kant holds that in reciprocity, or as he also calls it, community, "one thing is not subordinated, as effect, to another, as cause of its existence, but, simultaneously and reciprocally, is co-ordinated with it, as cause of the determination of the other" (B112). Further, in the third analogy, he writes that in reciprocity,

> Each substance . . . must therefore contain in itself the causality of certain determinations in the other substance, and at the same time the effects of the causality of that other; that is, the substance must stand, immediately or mediately, in dynamical community, if their coexistence is to be known in any possible experience (A212–B259).

Thus for Kant, the relation of reciprocity is one in which substances causally interact with each other. That is, each stands in the relation of both cause and effect to the other. Thus, in this relation they are equivalent. This relation of reciprocity is distinguished from causality in that the phenomenal appearance of the sequence of cause and effect is indifferent. That is, it makes no difference in what order one perceives each substance as the cause of an effect in the other.

Kant also briefly mentions the relation of reciprocity in his 'Table of Categories of Freedom' in the *Critique of Practical Reason*. There he cites it as "the relation of one person to the condition of another"[19] but unfortunately he does not discuss it any further. In this context, reciprocity pertains not to the relation among things as appearances, but rather to the relations among moral agents or persons. In this domain, the principle of causality is distinguished from that which holds among empirical phenomena. Here, he characterizes causality as having its determining ground in freedom, that is, in the agency of 'intelligible beings' or moral persons.

Certain features of Kant's concept of reciprocity are relevant to my proposed model. Like Kant, I see the relation as holding between independently existing entities. In my model, I take these entities to be agents, as Kant does in the Second Critique.

Again, I would adapt Kant's point about the equivalence of the relationship which each entity bears to the other in reciprocity to my model of reciprocal interaction among agents. I see this as a formal condition of reciprocity. However, it is not exhaustive of the full concept. I interpret the equivalence of the relation of reciprocity to consist in the fact that each agent in the relation takes his or her relation to the other to be equivalent. The equivalence is therefore constituted by the fact that each agent has the same understanding of the equivalence and that each freely agrees to act accordingly. Where agents are concerned, the relation is equivalent if they take it to be equivalent.

My model of reciprocity differs sharply from Kant's account in the First Critique in that I do not see the relation as involving in any sense reciprocal causality. That is, as I argued earlier, the actions of one agent with respect to another are not to be construed as causing the actions of the other. However, it should be mentioned that Kant sharply distinguishes agency or freedom from causality in nature and that therefore his conception of reciprocity among moral agents may be compatible in some ways with my own.

Another important philosophical account of reciprocity is provided by Hegel. I would like first to briefly consider the relation of my model to the account which he gives in *The Science of Logic* (at the end of the Doctrine of Essence). There Hegel sees the relation of reciprocity as going beyond causality, understood as an external or mechanical relation of cause and effect. He also sees reciprocity as going beyond Kant's notion of reciprocal causality, as a relation between two independent substances, each of which is the cause of changes in the other. Instead, he sees the deeper reality of reciprocity as the relation of cause to itself, namely as a self-relation of substance. This self-relation occurs through the mediation of substance's own activity, in which it posits itself also as effect. Thus Hegel writes, "*Reciprocity* is, therefore, only causality itself; cause not only *has* an effect, but in the effect it stands, *as cause*, in relation to itself .... In reciprocity

. . . the transition into an other is a reflection into itself."[20] In this sense, reciprocity for Hegel is really an internal reflexive relation, which he finally interprets as the relation of a totality to itself.

I do not accept Hegel's characterization of reciprocity as ultimately a form of self-relation, however self-differentiating such a relation may be. Thus it seems to me that Hegel ultimately eliminates the independence of the entities which stand in reciprocal relation and instead sees them as moments of a totality, understood as the Absolute. In contrast to this holistic view, I am proposing that reciprocal relations hold between independently existing agents and that therefore reciprocity cannot be taken as ultimately a form of reflexivity.

However, there is an important feature of Hegel's view which is relevant to my own model. And that is his comprehension of reciprocity as an internal relation and not as an external relation of reciprocal causality. I do not agree with Hegel's interpretation of internal relations as the total interdetermination of the nature and being of the entities within a totality. However, I would hold that reciprocal social interaction should be understood as an internal relation in the following sense: such an interaction is constituted by the shared understanding and free agreement of the agents who enter into it. The internality of the relation among such agents resides in the fact that both the understanding and the agreement are arrived at by the agents themselves with respect to each other and are not imposed from without nor conformed to as an external rule. Furthermore, the relation is also internal in that each agent is changed in the relation by his or her actions with respect to the other, but not by the causal action of one on the other. Thus on my view, reciprocity is an intentional relation, that is, it is constituted by the understanding and actions of the agents.

On the basis of the discussion of reciprocity thus far, I would propose the following as a working definition:

DEFINITION. A reciprocal relation is one in which each agent acts with respect to the other on the basis of a shared understanding, and a free agreement, to the effect that the actions of one with respect to the other are equivalent to the actions of the other with respect to the first.

It should be made clear that this definition in no way requires that the actions of both agents should be the same, but only that they be taken to be equivalent by the agents themselves.

I should now like to proceed to give a more concrete interpretation of this formal model through a consideration of the ways in which the concept of reciprocity has been developed in social theory. We may begin by referring again to Hegel, but in this case not to his logic, but to his social theory. In this context, Hegel gives two alternative analyses of reciprocity, both of which are important for my purposes. The first is his account of the so-called master-slave relation in the *Phenomenology of Mind* and the second is his account of abstract right and civil society in the *Philosophy of Right*.

In the master-slave dialectic, Hegel presents two features of social interaction which are important for my model. The first concerns his characterization of self-consciousness (which Hegel regards as constituting the very nature of a person). According to Hegel, the very existence of a self-consciousness requires that it be recognized. Thus he writes, "Self-consciousness exists in itself and for itself, in that, and by the fact that, it exists for another self-consciousness; that is to say, it *is* only by being acknowledged or 'recognized'."[21] But for Hegel, the very condition for being recognized by another is that this other be recognized as a self-consciousness. For only another self-consciousness would be able to recognize the first as what it is. This recognition by each of the other as a self-consciousness is clearly a reciprocal relation. Further, Hegel sees this reciprocal recognition in its full sense as the result of a process of development. This development is the condition for the full realization of an independent or free self-consciousness.

The relevance of Hegel's account here for my model of reciprocity lies first in his characterization of reciprocity as a relation which involves intentionality; that is, the relationship has as its necessary condition the consciousness or understanding of one agent by the other. However, I do not accept Hegel's radical intentionality, in which the very *existence* of the agent depends on being recognized by another. Rather, as I have said, I see this as a relation between independently existing entities. Hegel's account suggests an additional important feature of reciprocity which I will adopt. That is his understanding of reciprocal recognition as a condition for the full development of the agents as free. I shall call this fully developed form of reciprocity, in which it serves as a condition for freedom, intrinsic reciprocity or mutuality. I shall develop this conception further later on.

An equally significant feature of Hegel's account in the master-slave dialectic is his treatment of non-reciprocal social relations, that is, relations of domination and subordination. This inclusion of an analysis of non-reciprocity is very important in that it recognizes the insufficiency of reciprocity alone as a principle of social explanation. Hegel's account thus acknowledges in theory the social facts of coercion, oppression and domination and my model will follow his in this respect. Furthermore, he sees non-reciprocity and reciprocity as standing in a dynamic relation to each other such that non-reciprocity is seen as replaced by reciprocity through a process of development. Likewise, I shall take reciprocity not only as a principle of explanation but also as a norm with which one can criticize situations of domination as defective forms of social interaction.

It may be useful to remark parenthetically that not all forms of non-reciprocal relations are necessarily forms of domination or exploitation. One might mention as examples the cases of a free gift or benevolent actions done from duty without any expectation of benefit in return. In these cases, no reciprocal relation is involved. However, in what follows, where I consider non-reciprocal relations, my focus will be on domination and exploitation.

Hegel also characterizes the non-reciprocal relation between master and slave as one which yet requires a recognition by each of the nature of their relation. Thus Hegel sees as a requirement for domination that the slave acknowledges his subordination to the master. In this way, even in subordination, the agency of the slave is preserved. And indeed it is likewise crucial to my analysis, as I suggested earlier, that agency or the capacity to choose is not eliminated even under conditions of constraint, coercion or domination. Such agency of the oppressed is a necessary condition for the possibility of overcoming domination or of self-liberation.

Hegel's account of domination and subordination is also important in its analysis of this relation as one-sided recognition. That is, it is a relation in which the recognition of the slave by the master is not equivalent to the recognition of the master by the slave. Whereas the slave recognizes the master as an independent being, the master sees the slave as dependent and each sees himself in this way as well. I would define domination in a related way as a relation in which the actions of one agent are such as to control or delimit the direction or range of the actions of another. This relation is one of non-equivalence, since only one recognizes the other as a free agent.

An important difference between my account of domination and that which Hegel gives in his discussion of the master-slave relation concerns the origin and the means of continuing domination. For Hegel, domination has its source in and is based on force and violence, that is, on the threat to the life of the slave. I find this account inadequate as a model of social domination, since force and violence is only one of many instrumentalities of domination and it is not the main one. Rather, I would argue that the primary means of social domination is the control which an agent or a group of agents has over the conditions which another agent or group of agents requires for their actions. It is such control over objective conditions of agency which, at least on the social level, is the meaning of property. Where there is private property in the sense of private ownership of means of production, the social group which owns these means has control over the conditions which the other group without such property requires for their agency. We may note that such control over the conditions is one leading constituent of the meaning of social class.

An additional important feature of Hegel's account of domination and of the process of its overcoming is the relation which he establishes between social interaction, whether reciprocal or not, and the activity of transforming nature, that is, labor or work. Thus for Hegel, the domination by the master over the slave is mediated by the work of the slave on behalf of the master and through this very process of work, the slave achieves a recognition of himself as having mastery. Likewise, I would also want to claim that social interaction should be seen in its relation to causal agency, including here the work of transforming nature. The significance of this point for adequate explanation in the social sciences is that, in a complete account, the analysis of reciprocal or non-reciprocal social interaction must be supplemented by an account of causality as agency and also an account of the conditions for this agency and this interaction.

In the *Philosophy of Right*, Hegel develops a related model of reciprocity, but one whose emphasis is considerably different. In that work, Hegel is concerned with reciprocity as the foundation for the modern legal system and as the basis for economic exchange relations in what he calls civil society. He is no longer concerned with relations of domination and subordination, but rather with the relations among members of modern civil society and the state, who (in principle at least) reciprocally recognize each other as free and equal. In the section which is called Abstract Right, Hegel gives an analysis of property right as a reciprocal relation of persons who recognize each other as free and equal through the recognition of each other's right to own and alienate things. Further, he analyzes legal relations in terms of a reciprocity of rights and duties. Central to Hegel's account of civil society is his discussion of contract. He views it as an agreement in which persons freely and reciprocally determine the equivalence of whatever it is that they exchange, where this includes goods, services, obligations, etc. He uses this contract model as the basis for his analysis of economic exchange relations in the system of needs. In this system, each agent satisfies his own needs through satisfying the needs of another. They do this through the free agreement on the equivalence in value of the items which they exchange.

Based on Hegel's analysis here, I want to introduce a crucial aspect of my own model of reciprocity, namely, the specification of two delimited forms of reciprocity. I call these formal reciprocity and instrumental reciprocity. I would define formal reciprocity, in short, as the exchange of equivalents. It may be noted that this model of reciprocity has come to be the dominant one in the social sciences and has even been developed as a whole field of sociological theory and research named exchange theory (for example in the work of George Homans, Peter Blau and others). Instrumental reciprocity is a mode of formal reciprocity in which the exchange of equivalents is undertaken solely for the sake of each agent's gratification or for the satisfaction of each one's needs. Thus each agent recognizes the other only in terms of the benefits which the other can provide. The relationship becomes wholly instrumental, in that the equivalences are calculated on the basis of a cost-benefit analysis. In both formal reciprocity and in its specific mode as instrumental reciprocity, the reciprocal relation comes to be reduced to one of exchange. It may be noted here that even where this model has been used in the social sciences to describe and explain interactions which are not strictly economic, such interactions have been interpreted on the model of economic exchange.

The shortcomings of this model of the exchange of equivalents have been pointed out by Marx, in his further development of the concept of reciprocity. Especially in the *Grundrisse* and in *Capital*, Marx characterizes the system of exchange under capitalism in a way which is similar to Hegel's account. Yet he goes on to give a critical analysis of this system. First of all, Marx points out that the relations among the agents in the exchange appear only in the abstract form of the relations between the values of the commodities which they exchange. Thus the reciprocity in exchange is revealed to be merely formal. That is, the exchangers relate to each other not as concrete individuals with differentiated needs, capacities and interests, but rather only in terms of abstract equivalences expressed as monetary values. I will follow Marx here in that I will interpret instrumental reciprocity in its various modes as merely formal or abstract in this sense. By this I mean that in their relations, the significance of each agent to the other is reduced to only those properties, capacities and products of the agent's activity which figure in the specific relation of exchange.

The second critical dimension of Marx's analysis consists in his showing that the appearance of reciprocity in the system of exchange under capitalism hides the non-reciprocity in the system of production. As is well known, he argues that because capital controls the conditions of labor by ownership of the means of production, labor is not free not to contract with capital in order to gain means of subsistence. Thus, labor is constrained to exchange the capacity to work during a given period of time for means of subsistence. Therefore during this time capital gains disposition over this laboring activity and ownership of its products. This control over the direction and range of the agency of another is what I characterized earlier as domination; and it may be noted that in this case, domination has its origin and its instrumentality in the control over the objective conditions of agency. Beyond this, however, Marx's analysis reveals a form of non-reciprocal relations which he calls exploitation. This relation is based on the fact that through the exchange of wages for labor power in the market-place, the capitalist gains disposition over the value-creating activity of the laborer whereas the laborer gains the cost of reproducing his labor power. On Marx's analysis, the value created by the laborer during his or her working time is greater than the cost to the capitalist of the laborer's reproduction of his or her capacity to work. Marx calls this discrepancy surplus value. This is a measure of the exploitation which consists in the non-equivalence of their real economic relations. Specifically, the capitalist receives a portion of the value produced by the laborer without paying the equivalent. Thus Marx analyzed economic exploitation as a non-reciprocal relation.

I would also call attention to the fact that for Marx, the non-reciprocal relations of domination and exploitation in capitalism are internal relations. That is, on his view, capital and labor are defined relative to each other and each requires the other in order to be what it is. Furthermore, both capital and labor are changed in the relation of exploitation in that the power of capital grows relative to the diminution of the power of labor and the very character of each of them is changed historically through this relation. This point provides an example of the nature of reciprocal and non-reciprocal relations as internal relations which I proposed earlier.

An additional important point which I want to draw from the analyses of reciprocity given by both Hegel and Marx concerns the fact that reciprocal and non-reciprocal relations can be objectified. That is, such relations among agents can be embodied in and represented by institutions, social rules and practices. In this way, the forms of reciprocity and non-reciprocity may come to characterize objective social arrangements or social systems as a whole. On this basis, one may study social systems comparatively, with regard to the ways in which they embody and legitimate reciprocal or non-reciprocal relations. However, it would be wrong to see such systems or institutional arrangements as self-constituting or self-perpetuating. Rather, they are brought into being and maintained by agents.

The concept of reciprocity has been developed by a number of 20th century social theorists in ways that are relevant to my account. I shall briefly consider some of the specific points which they raise, before proceeding to a systematic review of my own model, as I have been developing it in this discussion.

Simmel's work on domination as a form of interaction is significant for my purposes because of its insistence on the agency or freedom even of the subordinate in the relation. Here he follows Hegel. His view is congruent with the point which I developed earlier, namely that agents do not cause or determine the actions of other agents. Thus Simmel writes,

> Even in the most oppressive and cruel cases of subordination, there is still a considerable measure of personal freedom .... [T]he super-subordination relationship destroys the subordinate's freedom only in the case of direct physical violation. In every other case, this relationship only demands a price for the realization of freedom, a price, to be sure, which we are not willing to pay.[22]

The phenomenological sociology of Alfred Schutz suggests an important respect in which the scope of the concept of reciprocity as I have thus far presented it needs to be broadened. In Volume I of his *Collected Papers*, Schutz describes what he calls the reciprocity of perspectives as a presupposition for everyday experience. This reciprocity refers to the fact that an individual takes for granted the interchangeability of his or her perspective with that of any other individual. That is, for Schutz, he or she assumes that any other individual would perceive or understand things in the same way if the other were in his or her place. Again, Schutz holds that it is also presupposed in everyday experience that the differences in perspectives which originate in individual biographical differences are irrelevant unless there is evidence to the contrary.[23] Indeed, Schutz sees reciprocity as an essential factor in the constitution of social relations in general. Thus he writes,

> The Thou-orientation is either one-sided or reciprocal. It is one-sided if I turn to you, but you ignore my presence. It is reciprocal if I am oriented to you, and you, in turn, take my existence into account. In that case a social relation becomes constituted. We shall define this relation formally as the 'pure' we-relation.[24]

Or again he writes," ... [T]he pure we-relation is constituted in the reciprocal Thou-orientation."[25]

Although I have serious reservations about Schutz's account, yet his discussion suggests an important addition to my own account of reciprocity. Namely, it emphasizes that an element of reciprocity is present in all social relationships. On my view, a dimension of reciprocity is implicit in social interaction even where there is no explicit shared understanding or explicit free agreement on equivalence. This dimension consists first, in the fact that each agent takes the other into account in the broad sense of acting with respect to the other, and second, that each implicitly acknowledges the other as agent. I would hold that this element of reciprocity is present even in non-reciprocal social relations. Although Schutz emphasizes the ubiquity of reciprocity in interaction, I do not agree with his view that reciprocity is a condition or presupposition for the possibility of social interaction in the sense that it is a ground in consciousness or in the form of an individual's experience for the constitution of such interaction. (I should perhaps add that it is not always clear how Schutz comes out on this question.) There is an additional significant feature of Schutz's account. He emphasizes that reciprocity functions in face-to-face interaction and this is a useful supplement to the previous emphasis in philosophy and in social theory on formal or contractual and institutional forms of reciprocal interaction.

Two recent views of reciprocity in social theory have been provided by Alvin Gouldner and Jurgen Habermas. In two important articles published in 1959 and 1960, Gouldner develops the concept of reciprocity as a norm, in part as a critique of functional theory in sociology. One of the most significant features of Gouldner's work is the centrality which he accords the concept of reciprocity as a working concept of explanation in the social sciences. He gives an analysis of reciprocity in which he shows that it differs from and goes beyond complementarity. Thus he writes, " ... complementarity connotes that one's rights are another's obligations, and *vice versa*. Reciprocity, however, connotes that *each* party has rights *and* duties."[26] He goes on to distinguish three forms of reciprocity: as the exchange of gratifications (or what I have characterized as instrumental reciprocity), as a folk belief and practice where reciprocity is taken as a 'fact of life', (what I would call customary reciprocity), and finally, reciprocity as a generalized moral norm. Although Gouldner articulates these important distinctions among different forms of reciprocity in social life, my criticism of his approach is that in all these forms, he interprets reciprocity on the model of exchange. Thus even in his account of reciprocity as a moral norm, he sees it as an obligation which people have to help those who have helped them and not to injure those who have helped them. Thus the moral norm itself is interpreted in terms of what I have called formal reciprocity, in this case interpreted as a reciprocal exchange of benefits. Gouldner leaves out of account intrinsic reciprocity or mutuality which I see as the highest stage in the development of reciprocal relations.

We might add that among recent theories of reciprocity, Gouldner's discussion is distinctive in the attention it calls to non-reciprocal social relations which he treats in terms of exploitation. However, this remains an undeveloped theme in these articles.

Jurgen Habermas discusses reciprocity primarily in the context of that form of social action which he calls communicative action. He takes reciprocity to be a decisive feature of what he calls the ideal speech situation. As such, reciprocity connotes "an unlimited interchangeability of dialogue roles (which) demands that no side be privileged in the performance of these roles."[27] Furthermore, inasmuch as, for Habermas, the ideal speech situation is presupposed in principle in all communication, reciprocity can be seen to play a role there. This recognition that some element of reciprocity is implicit in all communicative action is analogous to the view which I proposed to the effect that a dimension of reciprocity is present in all social interaction. In addition to this symmetry or reciprocity which Habermas attributes to the ideal speech situation, he also recognizes that existing empirical speech situations are constrained or distorted to various degrees and that therefore they are non-reciprocal. The ideal speech situation thus constitutes a norm for communicative action. Similarly, I have proposed that reciprocity constitutes a norm for social interaction, in terms of which one can criticize non-reciprocal relations of domination and their institutionalized forms.

One of the important differences between Habermas' view and what I am proposing here concerns his limitation of the scope of reciprocity to the sphere of communicative action. By contrast, I see it as applying to social interaction in general, which of course includes communication, but is not limited to it. Another major difference which I have with Habermas' conception of reciprocity is that it remains a concept of formal or abstract reciprocity in the following sense: within communicative action, the participants in the dialogue take each other into account only with respect to those features of their activity which are constitutive of the dialogue itself. That is, they are interchangeable in all of their dialogue roles. Undistorted discourse permits the presentation and discussion of all individual differences in views or interpretations, but its aim is to overcome them and replace them with a shared agreement. My criticism of this view is that it sees reciprocity as being realized in a situation of abstract universality, in which each agent is equivalent to any other, in the binding force of the consensus that they have rationally arrived at. I would claim that this view does not encourage the fullest differentiation of individuals or their freedom.

On the basis of the discussion thus far, I would like now to give a brief systematic presentation of the proposed model of reciprocity as it applies to social interaction. It may be useful to begin with a reiteration of the working definition of the general concept of reciprocity in social relations which I proposed earlier. A reciprocal social relation is defined as one in which each agent acts with respect to the other on the basis of a shared understanding, and a free agreement, to the effect that the actions of one with respect to the other are equivalent to the actions of the other with respect to the first. Such a reciprocal relation holds between human individuals conceived as agents who have the *capacity* to act freely. Further, the agents themselves define the equivalence of their reciprocal actions. Thus the equivalence is not to be understood as pre-existing or pre-given, but is rather constituted by their understanding and free agreement. It should be added that the equivalence which the agents determine need not be taken in narrow sense to refer to an equivalence between single actions, but may refer to sets of actions.

As I suggested earlier, some dimension of reciprocity is present in all social interaction even where it is only implicit and even when the relations are non-reciprocal. There is a minimal equivalence in every interaction insofar as each must recognize the other as agent, that is, as having the capacity to choose and act otherwise. Similarly, each must share some minimal understanding of the situation as a social relation. Thus even in the case of non-reciprocal relations, where there is non-equivalence, we have seen that there is an implicit acknowledgement by both, that the subordinate could do otherwise. If this were not the case, the notion of constraint or coercion would lose its meaning.

Another general feature which I introduced earlier is that both reciprocal and non-reciprocal social relations may be objectified or embodied in social institutions and in rules and norms of social practice, and thus may come to characterize social systems as a whole. Such institutional forms derive their force from the agents who act according to them.

The concept of reciprocity introduced here does not merely characterize social interactions descriptively, but in addition has a normative force. It can thus be used to criticize non-reciprocal relations both at the level of face-to-face interactions and institutional forms. I may suggest, without developing the point further, that the norm of reciprocity as I am presenting it here is closely related to the norm of justice.

We may proceed to a systematic sketch of the various senses of reciprocity which I discussed earlier. For in addition to the respect in which reciprocity is a general feature in all social interactions, it is developed in several distinctive forms in social life. These are: (1) customary reciprocity (2) formal reciprocity (3) instrumental reciprocity, as a specific kind of formal reciprocity and (4) concrete and intrinsic reciprocity or mutuality.

(1) By customary reciprocity I mean that form of social interaction in which the agents take each other into account in terms of tacit expectations of each other which they all share as part of their culture. The equivalence of the actions is established by everyone's following traditional or expected social practices. An example of this is the customary reciprocation of greetings in everyday encounters, such as one saying "Hello. How are you?" and the other responding "I'm fine, thanks. How are you?" However, some of the customs which govern such interactions specify actions which are highly differentiated according to status, position, role, etc. An example here is the ritual forms of greeting which acknowledge status difference honorifically such as "Your Honor" to a judge, "Mr. President" to the President, and "Your Majesty" to a king or queen. In such cases, the mode of reciprocal interaction serves to legitimate and perpetuate customary power relations and status differences which are themselves non-reciprocal relations. Thus because customary reciprocity is based on traditional, unreflective and uncritical acceptances of existing relations, it is especially prone to mask and even to strengthen non-reciprocal relations.

(2) Formal reciprocity, as presented above, is defined as the exchange of equivalents, where the equivalence of the actions and of the agents is established only in the exchange. This reciprocity is formal or abstract in that the concrete differences between what is exchanged and between the agents who enter into the exchange are irrelevant to their reciprocal interaction. Thus those accounts which emphasize the interchangeability of perspectives or the replaceability of any agent by any other in the relationship are formal or abstract in this sense. In this formal reciprocal relation, the agents take each other as equal within the scope of the exchange relation. But this equality is only abstract in that it is based on the reduction of the agents to their one-sided role of exchangers; in this role they are the same and in this sense equal. A characteristic example is that of an abstract moral norm in which all agents are obliged to respond to each other in the same way. Another example of formal reciprocity is the exchange of

commodities in the marketplace, in which the agents establish their equivalence as free and equal exchangers through their exchange of items which they agree on as equivalent. This is an example of the kind of formal reciprocity which I characterized as instrumental reciprocity.

(3) Instrumental reciprocity, as I defined it earlier, is a mode of formal reciprocity in which the exchange of equivalents is undertaken by each agent solely for the sake of each one's own gratification or for the satisfaction of his or her own needs. Each takes the other into account, but only as a means to his or her own ends.

Since both formal reciprocity and its specific mode as instrumental reciprocity are abstract or limited only to the equivalences established in exchange, they leave out of account all those features of agency and of social interaction which lie outside this sphere. In this way, these forms can exist side by side with non-reciprocal relations of domination and exploitation, and indeed can serve to mask the inequalities of social life under the appearance of formal equality. Moreover, in instrumental reciprocity, in particular, one agent acknowledges the other only as a means. But this falls short of the reciprocal recognition of the other as an agent in the full sense and also falls short of taking the other's needs into account. But this is precisely what would be required for a fully reciprocal relationship which I would call concrete and intrinsic reciprocity or mutuality.

(4) By full reciprocity or mutuality I mean a relation in which (a) each agent consciously recognizes the other as free, that is, as individualized and differentiated, and as capable of self-realization; and (b) each acts with respect to the other in ways which enhance the other's agency on the basis of a consideration of the other's needs and (c) both agents take such mutual enhancement of each other's agency as a conscious aim. On my view, reciprocity in this full sense of mutuality is a condition for freedom, where freedom means the self-realization of agents, expressed both in the development of their individual capacities and of their social relations. Furthermore, in this form, the equality of the agents in the relation may be called concrete, in the sense they are equally free in the development of their unique individuality and their social relationships.

Thus far I have presented the forms of reciprocity as stages in the development of a fully adequate concept of reciprocity, taken also as a norm. I would further propose that these stages may be interpreted also as stages in the historical development of forms of social interaction. Thus the first form, that of customary reciprocity, seems to me to have been the dominant form of reciprocal relations in pre-legal, pre-contractual and pre-capitalist societies, namely, societies in which the social bond is maintained primarily by tradition and custom. It may be noted that in most of these traditional societies, the customs often function to preserve and legitimate status differences and relations of domination. (I should mention in this connection that retribution or *lex talionis* is sometimes held to be the original and most primitive form of reciprocity. On my view, this principle might be characterized as negative reciprocity, since it is not an example of the norm as I have presented it, but is rather antithetical to it.)

Formal and instrumental reciprocity seem to me to be characteristic of the dominant forms of reciprocal interaction in capitalist societies, although they clearly have their roots in pre-capitalist exchange. These forms of reciprocity characterize the spheres of economic and political life, but also affect in some degree the modes of interaction in social, cultural and personal life. However, as merely formal, these modes of reciprocity mask non-reciprocal relations of domination and exploitation in social life, as I noted earlier.

Finally, I would propose that the fully developed form of reciprocity as mutuality provides a normative model for the development of new forms of social relations, in which freedom as the self-realization and self-development of social individuals is achieved through reciprocal recognition and mutual enhancement.

The final feature of this systematic review of my model of reciprocity concerns the analysis of domination. (As was remarked earlier, not all non-reciprocal relations are relations of domination or exploitation, but these are the most important for my purposes.) Domination may be defined as a relation in which the actions of one agent or group of agents are such as to control or delimit the direction or range of the actions of another agent or group of agents. I have argued that the primary means of such domination is control over the conditions which an agent requires for his or her actions. It is important to stress that these conditions may be either objective or subjective. Thus domination may proceed through psychological, as well as through economic or political means. Two of the main ways in which such control over the conditions of agency has been accomplished is by control over means of production, or again by the power to impose punitive consequences, whether social or psychological, on another for his or her actions.

It would be too extensive a task to critically examine in this paper the variety of concrete ways in which practicing social scientists have been using the concept of reciprocity. Among the key figures who should be considered in this regard are Mauss and Malinowski in anthropology; Parsons, Homans, Blau, Goffman and the ethnomethodologists in sociology; and Piaget and Laing among several others in psychology. Although I shall not treat these theorists here, it may be useful to suggest some general lines of criticism of their theories which emerge from a study of their work. (These criticisms which I shall only mention here are not intended to apply equally to all of them.)

First, with few exceptions, there is a pervasive tendency among these theorists to omit treatment of non-reciprocal relations of domination and exploitation from their systematic accounts and to focus on reciprocal relations. This is characteristic of functionalist explanations in sociology such as those of Parsons (a criticism which Gouldner has also made), as well as of those social anthropologists (like Fraser and and Malinowski) who presuppose reciprocity in the social exchange of gifts, services or other benefits. Where there is a consideration of status differences or power relations in such theories, these are regarded non-critically as expressed in and confirmed by the prevalent forms of reciprocity. Domination or exploitation, therefore, cannot be understood as systemic, that is, as a feature of the social system itself. There is a related tendency to subsume non-reciprocal relations under reciprocal relations by interpreting them in terms of so-called compensatory mechanisms or in terms of equilibrium sustaining adaptations. This approach is characteristic of such exchange theorists in sociology as Peter Blau, as well as of the work of Parsons. Here, again, the social facts of domination and exploitation are masked by interpreting them in the context of supposedly broader or more complex forms of reciprocity. Furthermore, these theorists tend to take a value-neutral approach to relations of power and domination in tacitly taking them to be legitimated by the very fact of their persistence.

A second criticism is that the model of reciprocity itself is reduced to formal or instrumental reciprocity especially in so-called exchange theory. Thus Blau, for example, writes, "'Social exchange,' as the term is used here, refers to voluntary actions of individuals that are motivated by the returns they are expected to bring and typically do in fact bring from others."[28]

In both of the ways criticized above, theories which adopt a model of reciprocity in place of a causal model nevertheless may retain an ideological dimension analogous to that which I described above in connection with the causal models of explanation. In their uncritical attitude towards social relations of domination and exploitation (either by exclusion or accommodation of such relations in their theory), such theories contribute to the legitimation of such forms of non-reciprocal interaction. Furthermore, where reciprocity is reduced to an instrumental relation, the particular characteristics of social relations under capitalism are proposed as constituting the nature of social relations in general.

A third criticism of prevalent models of reciprocity in the social sciences is that with few exceptions they are ahistorical. This is, they do not see either the forms of reciprocity or the forms of domination as historically changing. A final criticism of most of these theories is that they take the model of reciprocity to be merely a descriptive one, and do not see that it also has a normative and critical force. This is the case even where there is a description of reciprocity as a norm for a given form of society or of social interaction, as for example in Malinowski, Blau or Piaget. In these cases, the norms are merely described and are not themselves evaluated critically.

In contrast to these views, the model which I have proposed sees reciprocity as both a norm and a principle of explanation in the social sciences. These aspects need to be methodologically integrated so that such explanations are not merely conceived as value-neutral and descriptive but also as critical.

How, then, does reciprocity serve as a model for explanation in the social sciences and what is its normative force? Here I will only briefly sketch some ways in which the model may be applied in social scientific research. The model of reciprocity differs sharply from causal models of explanation which, as I argued in the first part of this paper, are incompatible with the freedom of agents. By contrast, the model of reciprocity takes the capacity for freedom as a fundamental characteristic of agents and thus as a presupposition of any explanation of social interaction. Furthermore, this model differs from both methodological individualism and methodological holism as approaches to explanation in the social sciences. Holism gives primacy to social structures and institutions and sees the actions of individuals as determined by their role within such structures. It thereby omits an account of the agency of individuals. By contrast, individualist approaches reduce social phenomena to aggregates of individual actions, in which these actions are explained in terms of the motives or drives of individual egos. It thereby omits an account of objective social structures and the social character of individuals. The model of reciprocity differs from both of these approaches in that it takes as its fundamental entities social individuals who constitute the forms and structures of social life by their activity. This model therefore leads one to focus both on interactions at a personal level and on institutionalized systems of interaction within a social whole. In studying the social whole, this model sees it as ultimately constituted by the activities of agents. And in studying interpersonal relations, this model also sees them in relation to social and cultural contexts.

Reciprocity serves as a principle of explanation in the social sciences by providing a distinctive class of reasons that agents have for their actions, namely, that they seek to establish equivalences in their interactions with other agents. In explaining actions, the social scientist would see them not simply in terms of the individual agent's needs and purposes but also in the context of the interaction among agents. Thus the model of reciprocity takes the unit of investigation to be an interaction. In this way, reciprocity as a principle of explanation supplements intentionalistic accounts which focus on the purposes of individuals. Nonetheless, reciprocity implies that the interactions of agents are intentional in the sense that the reciprocal actions are based on the conscious purposes and free agreements of the agents. Thus, explanation in terms of reciprocity, like all explanation of action, requires that the actions be understood in terms of the reasons that the agents have for their actions. Further, since in reciprocal relations, such reasons are based on the self-understanding (both individual and shared) which agents have of themselves in the relation, any explanation in terms of reciprocity would have to take this self-understanding into account.

In the explanation of actions and interactions, the social scientist would therefore investigate empirically the shared understandings and expectations which the agents have, their modes of reaching agreement and the nature of the equivalences which they establish in their relations. The shared understandings of the subjects of such an inquiry might include not only matters of personal agreement, but also common cultural or moral norms, explicit laws or rules of conduct, as well as various types of background knowledge including a common language and culture. The modes of reaching agreement might include negotiation and contract, mediation or arbitration by a third party, assignment of the power of decision to an external authority, or more commonly by the less formal means of rational discourse, or appeals to feeling or intuition, among others. Furthermore, while the nature of the equivalences established in reciprocal relations may in general be characterized either as the same action on the part of each agent or as different actions taken to be equivalent by each of them, yet the specific actions taken as the same or equivalent would vary from case to case. It would be of interest to study what specific equivalences are established in concrete cases and what the motivations or reasons of the agents are for taking such actions as equivalent.

It would also be important to distinguish the various types of reciprocity which are evident in different social relations. Thus the shared understandings, methods of reaching agreement and the nature of the equivalents are different in the case of traditional or customary reciprocity (e.g. the giving and receiving of gifts or greetings) from what they are in the case of formal and instrumental reciprocity (e.g. market exchange, legal contract). And both of these differ from the case of intrinsic reciprocity or mutuality (e.g. love or friendship).

However, it is also crucial for social science to demarcate and to study non-reciprocal forms of social relations and in particular those of domination and exploitation. Here, one would be concerned to discover those cases which exhibit a lack of shared understanding – whether misunderstanding or conflicting understandings, or in which there is a lack of free agreement on the part of one of the agents, that is, where one agent is coerced by another. Further, in these cases, one would study the inequivalences or inequalities in the actions of the agents with respect to each other. Here too, it would be important to discover the motives or reasons which agents have for dominating others or for abiding subordination. In addition, one might attempt a genetic or historical account of the ways in which such relations of domination and subordination have come to be established.

Some of the most interesting cases for social scientific inquiry are those in which reciprocal and non-reciprocal social relations exist side by side; and indeed those in which the forms of reciprocal relations serve to mask the non-reciprocal relations of domination or exploitation. For example, in face-to-face interaction in the work place, relations which appear reciprocal may in fact hide non-reciprocal or hierarchical relations. Thus a division of tasks among managerial and non-managerial personnel may be presented as if it were one in which all are making an equal contribution, each in his or her own way. However, the equivalence is belied by the fact that the manager can hire or fire the workers, but not vice versa, and that the workers ordinarily do not have a say in determining the direction of the work process. Or again, at the level of social or economic life as a whole, one would be concerned to determine whether systems of non-reciprocity coexist with systems of reciprocity and if so, what the relations are between them. Thus for example, the reciprocity in the system of economic exchange has been studied in relation to the non-reciprocal relations of exploitation in the system of production. In addition, one might consider the formal equality of the law in relation to concrete and systematic inequalities in its various applications. Furthermore, on this model the social scientist would study these various systems of reciprocity and non-reciprocity in their historical development. Moreover, he or she would try to relate such systems of institutional interaction to the modes of personal interaction. In addition, the social scientist would study the dynamic relation between reciprocal and non-reciprocal forms of interaction, whether personal or institutional, and would attempt to discern the possibilities for the transformation of non-reciprocal relations into reciprocal ones.

It is evident from these last considerations that the model of reciprocity is not only a descriptive and explanatory instrument but that it is also normative and critical. As I have shown, the normative force of the concept of reciprocity lies in the fact that reciprocity, especially in its most elaborated form as mutuality, is a condition for the full development of the freedom of agents. Therefore, in the interest of promoting freedom, the social scientist should be critical of the non-reciprocal social relations of domination and exploitation in which the freedom of agents is constrained. Further, he or she should make clear what the forms of reciprocity are that exist in various social or personal spheres and should seek to show the ways in which these reciprocal social relations may be developed into those of mutuality.

The model of reciprocity, as I have developed it, is intended to apply to social relations in face-to-face situations and in social systems as a whole. I suggest that it would be interesting to apply the model of reciprocity to such face-to-face situations as those of the workplace, relations between men and women, greetings and partings, gestures, and speech and written communication among individuals. In the domain of social institutions and entire social systems, one might use the model to analyze such cases as the relations among social classes, among interest groups, and among nations; to study such contexts as negotiations, punishment and rights and claims in the law, economic exchange and distribution; and to analyze such general activities as language use and other forms of cultural communication. In these cases, one would study the nature of the equivalences and how they are agreed upon, as well as the forms of domination and how they are established. Beyond this, it would be important to discover whether there are any fundamental forms of reciprocity or domination that are characteristic of some of the personal or institutional relations in a given social system or in a given historical period.

A final remark, by way of summary, is appropriate here: namely, that as a principle of explanation, reciprocity is not enough. Rather, as I have claimed, a complete explanation in the social sciences requires in addition an account of the intentionality of agents, the causality of their agency in the sense of their productive and creative modes of activity, as well as an account of the conditions of their agency. Such an approach to explanation in the social sciences would seek to understand the connections between social relations and human agency and between mutuality and human freedom.

Stevens Institute of Technology

NOTES

* Originally presented at The Boston Colloquium for the Philosophy of Science, Boston University, March 15, 1977. I would like to thank Hugh Lacey, Joseph Margolis and Marx Wartofsky for their very helpful comments and criticisms of earlier versions of this paper. I would also like to thank Ken Gergen, Otto Marx, Jonathan Moreno and Barry Schwartz for discussions which served to clarify various issues in the paper.

[1] M. Mauss, *The Gift*, tr. by I. Cunnison (Free Press, Glencoe, Ill., 1954).

[2] B. Malinowski, *Crime and Custom in Savage Society* (Routledge and Kegan Paul, London, 1961).

[3] G. H. Mead, *Mind, Self and Society from the Standpoint of a Social Behaviorist* (Univ. of Chicago Press, Chicago, 1934).

[4] G. Simmel, 'Superordination and Subordination,' in *The Sociology of George Simmel*, ed. and tr. by K. Wolff (Free Press, Glencoe, Ill., 1950). Part III, pp. 181–303.

[5] A. Schutz, 'Common Sense and the Scientific Interpretation of Human Action' and 'Concept and Theory Formation in the Social Sciences' in *Collected Papers* (Martinus Nijhoff, The Hague, 1971), Vol. I, pp. 3–47, 48–66.

A. Schutz, 'Social Reality within Reach of Direct Experience,' in *Collected Papers*, Vol. II, pp. 22–36.

A. Schutz, *On Phenomenology and Social Relations*, ed. by H. R. Wagner (Univ. of Chicago Press, Chicago, 1970).

A. Schutz, *The Phenomenology of the Social World* (Northwestern Univ. Press, Evanston, 1967).

[6] J. Piaget, *The Moral Judgment of the Child* (Free Press, New York, 1965).

[7] T. Parsons, *The Structure of Social Action* (Free Press, Glencoe, Ill., 1964).

[8] G. C. Homans, *The Nature of Social Science* (Harcourt, Brace and World, New York, 1967).

G. C. Homans, *Social Behavior: Its Elementary Forms* (Harcourt Brace Jovanovich, New York, 1961).

[9] P. Blau, *The Dynamics of Bureaucracy* (Univ. of Chicago Press, Chicago, 1955).

P. Blau, *Exchange and Power in Social Life* (J. Wiley and Sons, New York, 1964).

P. Blau, *On the Nature of Organizations* (Wiley, New York, 1974).

[10] A. Gouldner, *The Coming Crisis of Western Sociology* (Basic Books, New York, 1970).

A. Gouldner, 'The Norm of Reciprocity: A Preliminary Statement,' *American Sociological Review* **25** (1960), 161–178.

A. Gouldner, 'Reciprocity and Autonomy in Functional Theory' in L. Gross (ed.), *Symposium on Sociological Theory* (Row, Peterson, Evanston, 1959).

[11] E. Goffman, *Behavior in Public Places* (Free Press, New York, 1963).

E. Goffman, *Encounters* (Bobbs-Merrill, Indianapolis, 1961).

E. Goffman, *Presentation of Self in Everyday Life* (Doubleday, Garden City, N. Y., 1959).

E. Goffman, *Relations in Public* (Basic Books, New York, 1971).

E. Goffman, *Strategic Interaction* (Univ. of Pennsylvania Press, Philadelphia, 1971).

[12] E. Husserl, 'Philosophy as Rigorous Science,' in *Phenomenology and the Crisis of Philosophy*, tr. by Q. Lauer (Harper & Row, New York, 1965), p. 81.

[13] Although I share the phenomenologists' emphasis on the character of human action as free, yet I believe that their conception of human activity is inadequate and that their account of freedom is insufficient. It may be useful to sketch briefly some main features of these phenomenological views of freedom.

By contrast to the causal account of action, Husserl proposes an account in terms of the intentionality of consciousness as a presupposition of all empirical knowing and doing. This intentionality refers to the directedness of acts of consciousness towards objects, where this relation is one between a meaning constituting act and the thing meant. These constitutive acts of consciousness which give rise to the multiplicity of objects of consciousness themselves presuppose what Husserl, in his earlier works, calls 'absolute subjectivity.' This absolute subjectivity is taken by Husserl to be self-generating or self-originating activity and thus not the product or the effect of any external cause. (This earlier concept seems to be subsumed in the notion of a transcendental ego, which Husserl develops in his later works, e.g., *Cartesian Meditations*).

Heidegger, Sartre and Merleau-Ponty further develop this notion of the self-activity of the subject, not simply in the context of reflective consciousness but also in the contexts of human action and practice. In their various analyses, they put emphasis on the character of action as fundamentally free. Thus in *Being and Time*, Heidegger conceives of the fundamental structure of human activity as purposive and future-directed. Dasein or human being constitutes or structures situations in terms of what Heidegger calls its projects. He further characterizes Dasein as being essentially the sort of being which projects possibilities for itself and which determines itself to be the sort of being it is by choosing these possibilities. Similarly, for Sartre, consciousness, or what he calls the for-itself, is free in that it constantly "puts itself in question" or chooses directions for itself.

[14] See C. Gould, *Marx's Social Ontology: Individuality and Community in Marx's Theory of Social Reality* (The MIT Press, Cambridge, Mass., 1978), Chapter 3.

[15] It may be useful to compare and contrast this conception of the causality of agency with the phenomenological conception of constitution. For Husserl, the notion of activity or of agency is interpreted only in terms of acts of consciousness. All other modes of action are 'bracketed' as pertaining to what he calls the 'natural standpoint.' The acts of consciousness constitute the meanings of objects (and in the late work, constitute the very Being of objects as their meanings). Husserl describes this constituting activity as an activity of synthesis, in which consciousness establishes the identity of an object through its multifarious appearances. In this concept of constitution,

there is implicit the notion of form-giving activity, but it is understood only as the creation of meanings within consciousness and has no sense of causal efficacy. By contrast, I would interpret constitution as the causal activity of agents, that is, as practical and productive activity, in which agents work on and transform objects.

In Heidegger and Sartre, as against Husserl, activity and the constitution of meanings is taken to be in the world and not in consciousness alone. This difference derives in large part from their rejection of Husserl's theory of a transcendental ego. Thus, for Heidegger, Dasein or human being is characterized as Being-in-the-world. Similarly for Sartre (in his work *The Transcendence of the Ego*), the ego is seen as out there, in the world, and as constituted by the conscious activity of a subject. Heidegger and Sartre thus offer a more practical account of constitution in construing it in terms of the purposes and intentions of a subject existing in concrete situations. Yet, neither of them sees this activity as a causal process, in which an agent works on and transforms given objective conditions.

It may be seen that Marx's view of constitution and of agency differs from that of the phenomenologists in that he sees the activity of agents as causal and interprets such activity as the transformation of objective conditions in accordance with the agent's purposes. In this respect, my view has much in common with that of Marx. Like him, I would characterize human action with respect to objects in terms of productive causality and would see this causality as itself intentional in the sense that the activity is undertaken for the sake of realizing the conscious purposes of agents. Marx introduces a conception of the activity of constitution as an activity of laboring which is causally effective and gives meaning to things by transforming a given world in accordance with the subject's purposes. For Marx, labor is a process of objectification in which an agent forms objects which embody his or her intentions or purposes and in doing so also forms him- or herself. In this process, objects are not to be understood simply as givens, but are rather to be regarded as produced by the agent's activity. However, the agent does not constitute objects out of nothing. Rather, the agent of subject works on something which is given to him or her, as external or other.

Thus, for Marx, "labor is purposeful activity" (*Grundrisse*, tr. by M. Nicolaus (Vintage, New York, 1973), p. 311). In order to realize his or her purposes, the individual in the activity of labor gives form to objects such that these purposes are realized. Thus Marx calls labor a "form-giving activity" (*Grundrisse*, p. 301). Further, insofar as the objects created by this activity realize the purposes or satisfy the needs of the agent, these objects come to have a use or value for the subject. Accordingly, Marx characterizes labor as 'value-positing activity.' But perhaps the most significant feature which Marx attributes to the activity of labor is that it is productive activity in the sense of being creative activity, that is of creating new objects. These objects in turn constitute the objective conditions for subsequent laboring activity.

[16] My view here differs from that characteristic of most phenomenologists. I would claim that Husserl, Heidegger and Sartre among others do not sufficiently take into account the objective conditions of human agency including both the natural world and the domain of social interaction. Most problematically, in its focus on intentions, this view does not take into account the objective social and historical circumstances which condition action. Insofar as the phenomenologists do consider the social circumstances of action, such circumstances are regarded as the horizon or conscious

framework for one's activity, and thus from my standpoint do not have the requisite objectivity or sociality.

[17] K. Marx, *Grundrisse*, p. 161.

[18] *Ibid*., p. 157.

[19] I. Kant, *Critique of Practical Reason*, tr. by L. W. Beck (Bobbs-Merrill, Indianapolis, 1956), p. 69.

[20] G. W. F. Hegel, *The Science of Logic*, tr. by A. V. Miller (George Allen and Unwin, London, 1969), p. 570.

[21] G. W. F. Hegel, *The Phenomenology of Mind*, tr. by J. B. Baillie (George Allen and Unwin, London, 1910), p. 229.

[22] G. Simmel, *The Sociology of George Simmel*, p. 182.

[23] A. Schutz, *Collected Papers*, Vol. I, pp. 10–11.

[24] A. Schutz, *Collected Papers*, Vol. II, pp. 24–25.

[25] *Ibid*, p. 27.

[26] A. Gouldner, 'The Norm of Reciprocity,' p. 169.

[27] J. Habermas, 'Toward a Theory of Communicative Competence,' in H. P. Dreitzel (ed.), *Recent Sociology*, no. 2, p. 143.

[28] P. Blau, *Exchange and Power in Social Life*, p. 91.

MARJORIE GRENE

EMPIRICISM AND THE PHILOSOPHY OF SCIENCE OR, *n* DOGMAS OF EMPIRICISM

Thomas Reid remarked of what he believed to have been Descartes' achievement: "To throw off the prejudices of education, and to create a system of nature, totally different from that which had subdued the understanding of mankind, and kept it in subjection for so many centuries, required an uncommon force of mind."[1] So it is now, in turn, with the Cartesian heritage. Part of the difficulty, for us, of casting off that albatross comes, I believe, from the special empiricist version of the Cartesian view that has dominated philosophy in the Anglo-American tradition. This is true of philosophy in general and *a fortiori* of philosophy of science in particular. I shall try to illustrate this thesis by pointing to tenets characteristic of each of the major British empiricists that have, in my view, contributed to the impasse in which twentieth century thought has found itself. A necessary condition for the empiricists' missteps, of course, was the Cartesian alternative itself, between pure intellective mind on the one hand and bare inert monolithic matter on the other; but that singularly abstract and unlikely division acquires additional malignity in the still more abstract and unlikely caricature of experience characteristic of empiricist thought, at least in its British version, which is all I shall be looking at. Presumably there are equivalent absurdities to be found in such writers as Condillac; and philosophy of science, of course, had one of its chief origins in continental positivism, an analogue of our empiricist tradition. The lessons I am looking for can be gleaned, however, from the empiricist writers most familiar to English speakers, and, if only out of ignorance, I shall confine myself to them. In each case I shall also try to indicate very roughly what a more adequate alternative to the errors in question might look like. Mind you, I am *not* saying that Locke, Berkeley and Hume did all this on their own, but only that we can get a kind of profile of the philosophy back of philosophy of science as it has looked until recently if we glance at some of the doctrines of some of its forebears. What this backward glance produces will be, I fear, rather a jeremiad than an argument. That's the trouble with being an historicist: our philosophical situation appears to me so desperate, and I find myself (lacking the uncommon force of mind aforementioned) so entirely powerless to alter that situation, that by now all I can do is scream.

R. S. Cohen and M. W. Wartofsky (eds.), Epistemology, Methodology and the Social Sciences, 89–106.

So here goes. To begin with, Locke. What most plainly characterizes Locke's position with respect to scientific inquiry is what we may call his thin realism, the same general conception, I'm afraid, around which most controversy about so-called scientific realism has revolved. This realism is in turn associated with other characteristic features of Locke's doctrine that have also plagued our thinking about science. Let me try, schematically, to put these doctrines in place. First, Locke like Descartes before him and Berkeley after him lived in God's world. We seem to believe we can drop this supernatural feature and keep the rest if we like. But can we? Descartes at least recognized that a thinking, in particular a mathematicizing, mind could reach out to an inert nature only with God's complicity. What remains, outside any relation to God's veracity and benevolence, is intrinsically incoherent. No good could come of it, as no good has. Nevertheless, let us look, as we have usually done, at the features of Locke's epistemologized cosmology, minus its necessary condition in a divine source. With Boylean and Newtonian additions, Locke of course retains belief in the featureless one-level extended stuff of the Cartesian dichotomy. Ideas, however, he demotes from acts of a clear and attentive mind to a sort of grocer's stock. Although the mind is still conceived as active, its contents are dumped into it either by sense or by an introspection of its own activities. Even God has a tough time making such a situation coherent. There is this stock of ideas, 'mental' because inherent in my mind; there is that spread of particles, equally inert, of whose existence our hero Newton and our friend Boyle have persuaded us. How even God can suitably relate them to one another, remains a mystery. For a Scottish divine like Reid, that's fine: why *should* we understand? Mind-body interaction, and therefore mind's knowledge of nature, is of course a perfect mystery. But if we look at the Lockean scene from a more secular perspective, its incoherence and instability are what most strike us – the very instability and incoherence, I believe, that have characterized philosophy of science throughout its short official history. On the one hand, Locke tells us over and over, we are conversant only with our ideas, and knowledge is the perception of agreement and disagreement among ideas. On this side, Locke's uneasy representationalism is bound to issue, as it did in Berkeley, in a phenomenalist position. This way lie protocol sentences, or Wigner's boast that modern physics reports only the physicist's sensations. As if there could be science except in a real world where real people are puzzled, through wanting to understand, with whatever abstract sophistication, how something in the real world really works. But Locke's representative theory leans equally toward another issue:

representations must represent *some*thing, and of course *what* they represent will be that jolly corpuscularian matter substituted by good Royal Society members for the unintelligible forms or species or substrates of the schools. On the one side, then, there is the inventory of ideas, on the other, the objective of the new corpuscular philosophy: that is, least particles of matter, with their shape, size, impenetrability and texture: the last, a lovely Boylean addition. How, except by its texture, Boyle asked, do we tell sound from spoiled cheese? At least there is a link there with the perceived world. In the main, however, the very virtue of corpuscular realism seems to be the imperceptibility of its little bits, its occultness, indeed. The sensible, I suppose, keeps changing; only in its insensible parts do we find the solidity we seek. "God probably formed matter out of hard solid impenetrable particles." To Locke and other Boyleans this seemed to be a bottom line any reasonable man could rely on. Actually, although I have just ventured to give a kind of reason, I have never really understood quite why this should be so. What *is* so explanatory about least parts? Yet so it has seemed recurrently since Democritus, and so it seemed to Locke. He is caught, therefore, irretrievably between his version of the New Way of Ideas on one side and the New Corpuscular Philosophy on the other. He believes in both although he cannot consistently believe in both. Those who stress his serious commitment to the corpuscularian creed are quite correct; but so are those who stress his implicit phenomenalism. Look at the list of degrees of knowledge, for example. Knowledge has been defined as the perception of agreement and disagreement among ideas. This may be intuitive or demonstrative. But what is the third kind, sensitive knowledge? It must consist in the agreement or disagreement, not of ideas, but of ideas and something non-ideal that causes them. But that is not agreement among ideas at all. Even more glaringly inconsistent: in the classification of relations between ideas we find real existence, again not a relation between ideas at all, but between ideas and something wholly, categorically different, to which, by Locke's own account, we could have no access. If we are conversant only with our ideas, in other words, we have no knowledge of real existence; if we have knowledge of real existence, we are not conversant solely with our ideas. Further – and here is the real trouble – what, in the last analysis, could the connection be between the supermarket of ideas and the austere corpuscularian events of the non-ideal material world? God only knows, and orphaned of His protection, philosophy of science has swung forlornly between a phenomenalist account of hypotheses tying together sense-datum type reports with at best pragmatic (and therefore uncomprehending) interconnections, and the nothing-buttery of atomism, subatomism, quarkism or what you will. To put the alternative in more outspoken ontological terms: on one side there is a meagre mentalism, an aggregate of passive subjective states of the sort of which, Richard Rorty has argued, incorrigibility is the mark.[2] Now this has in my view very little indeed to do with mind in any recognizable form. It has to do with nasty pains and jolly flavors and possibly little after-images and such; but from minding as we live and know it, it seems remote indeed. Yet this ghost of mind has served in our tradition as the chief surrogate for the mental. And over against the ghost is the vaunted machinery of so-called matter: equally thin and disembodied, a tattered substitute for the flesh and blood nature into which we were born, in which we live and out of which we will breathe our last. If not the minimal, subjective mental, then the equally minimal, physicalistic material is all we are offered. Fodor's 'Disunity of Science' paper, for example, seems, if I understand it at all, to have to do with a problem about the laws of psychology on the one hand and of fundamental physics on the other. But why on earth leap from subjectivity to fundamental physics, when, plainly, mental life is a form of *life*, expressing not only its own structures (which are not just subjectivistic either), but those of the complex, hierarchically organized life-styles that support it? But no: in the Lockean tradition, which willy-nilly we have inherited, it is only the subjective flow of alleged bits of mentality or the one-level collisions of least bits of stuff that are allowed to count.

And what can unite these two abstractions? A very thin thread indeed, stretched like one spider's filament from a presumed observation point to a presumed reticulum of corpuscles in turn suspended in absolute space beneath. Why ever should writers flaunting these improbable hypotheses laugh at the story of the earth as resting on a giant elephant resting on a giant tortoise resting on one knows not what? There was more solidity to that image, at any rate! Yet this same attenuated realism is the only one philosophy of science (again insofar as I understand the arguments in question) has ventured to claim. The problem seems to be, precisely as it was for Locke, how we can get out of our mentalistic, subjective skins to make contact with a completely alien reality, a reality that *must*, thanks to our obeisance to the newest 'new corpuscular philosophy,' exhaust itself in the regularities prescribed by fundamental physics. But how under the sun did the latter-day saints of science achieve their acquaintance with the fundamentalities of nature, which we so humbly accept from them? What blind man's probe had they that they could grope their way into secret hiding places even the sighted couldn't see? Such an unintelligible contact of mere mind with mere matter was all very well when God guaranteed the harmony of subjective and material. Even so, enough controversy raged about it. For us, however, the question of realism has become a triple pseudo-problem: how to connect a pseudo-mental series of subjective states with a pseudo-nature by some hypothesized connective. Good lord, where *are* we?

I warned you this would be a jeremiad. Where are we, indeed? At one level, at a dead end, philosophically, engaged in Quixotic (or Pickwickian?) battles surely no less absurd than those despised by the founders of our now defunct tradition. But that's only philosophy. Really, we are where every one always is: in the doubly real, natural and human world, of some aspects of whose history each of us forms one fragmentary yet significant expression. This alternative and much more plausible account of where we really are lacks a ready professional name. The name a few of us know it by is that of being-in-a-world, in the Merleau-Pontyan meaning of that phrase. And I do believe by now that in every aspect of philosophical reflection only a really radical fresh starting point on the foundation of some such conception will allow us *any* fruitful discussion of any philosophical issue whatsoever. Maybe for now, however, I can call this new beginning 'comprehensive realism,' meaning by this the proposition that human beings like other animals are comprehended within a real living environment, and though that's not the primary point here, within a human social world, or better a family of human social worlds, as well. Although Heidegger missed almost entirely, indeed despised, the containment of human existence in the living world, he was right to start his description of *Dasein* with an account of *In-Sein*. In contrast, Cartesian *res cogitans* could be only external to its other, *res extensa*. In other words, indwelling, as Michael Polanyi called it, is chronically impossible for a Cartesian-Lockean mind. Ideas are captured by the supposed subjectivity of the mind, primary qualities of external things by the unknown substrate in which God (or physics) has placed them, and never the twain shall meet. No wonder many modern thinkers would like to abandon the ghost altogether for the machine – but, again, how do we know about the machine? From what other ghosts have told us? Ghosts don't communicate. What nonsense! Do let us try, even though we are, alas, philosophers, English-speaking and therefore by inheritance empiricist philosophers, to jump off this merry-go-round and return for the starting point of our reflections to where we are: to the living-times-human world in which people orient themselves, and learn all sorts of practices from baseball to the sciences, into which again they enter as they learn. That science itself is a family of practices, not just one, as empiricism saw it, is, again, not to the point here. The point here is just that we start, and finish, as real beings in a real natural and historical world, in which we find our insights and our problems, a world that in our case includes scientific institutions, scientific beliefs, scientific languages, scientific methodologies, which some of us have, and take, the opportunity to assimilate, and even to modify, once we know how, in partly novel ways. There is no problem of scientific realism; there are sciences which originate and develop within a real world. Thus comprehensive realism in turn generates what I suppose one might call an ecological epistemology – to be sharply distinguished, however, from that singular red herring, evolutionary epistemology. (I have criticized the latter on other occasions; it should be firmly put aside here. What I am trying to refer you to now is the notion of the real containment of real living persons in their real natural and cultural surroundings. On this view, we are not, or certainly not only, aggregrates of least bits resulting from the efforts of those bits [whatever that can mean!] to perpetuate themselves.)

Comprehensive realism, then, entails *being-in* as a fundamental category, or, in Heideggerian terms, an existential rather than a category. Being-in-a-world, moreover, is neither mentalistic like Lockean ideas nor physicalistic in terms of either an older or a newer corpuscularian philosophy. It is multi-dimensional: including always bodily, not only fundamental-physical, or even biochemical, realities. As the German language can put it, it includes *Leiblichkeit*, not only *Körper*, lived bodiliness, not only the spread-outness of inertial matter or even the forces that, unintelligibly, have been found impressed on it. And at the same time, being-in-a-world always includes, for us, a dimension or many dimensions of mentality: perception, thought, passion – all resonances of our bodily being, and yet that bodily being also resonates to our mentality. As Merleau-Ponty put it in the *Structure of Behavior*, the physical, living and human orders are, in our existence, all three always compresent – and neither is any one of them necessarily one simple dimension. It is rather, in a borrowed phrase of Merleau-Ponty's, "patterned mixed-upness" that characterizes the real within which we find ourselves. Call this multi-dimensionality if you prefer.

So much for our debt to Locke. There are of course other confusions in the *Essay* besides its central incoherence. For example, Locke plainly didn't know the difference between particulars and universals. I shall return to that problem very briefly in connection with the nominalism of Hume and Berkeley. Meantime Berkeley's chief contribution to my catalogue of intellectual disasters must be acknowledged. Granted, the dear Bishop, then future Bishop, produced in the *De motu* some cogent criticisms of Newtonian causality, as well as anticipating some falsehoods about the purely conventional status of theoretical entities; I don't know how influential that work was. His chief influence, however, in two disciplines, philosophy and (once it began) experimental psychology, stems from his love of what were called "minimum sensibles" and the misguided theory of perception he constructed with their help. Associated with this bizarre doctrine are the distinction between sensation and perception, and with that again the judgmental view of perception. True, the latter was already expressed by Descartes, but at least he let us 'see' hats and cloaks, not only inverted ragged two-dimensional colored patches. It was Berkeley in the *New Theory of Vision* who really set the whole thing up. Lockean simple ideas have now become *minima*, the least bits apprehended by any one sense, by means of which, or by means of aggregates of which, we manage to predict other bits to come. By our time, indeed, the notion of perception as judgmental has become so commonplace that we fail to notice the absurdities into which it leads us. But if you would just reread the *New Theory of Vision*, it might jar you out of your complacency. Whether I look at the wall in front of me, Berkeley holds, or out the window at the valley, or walk up the hills and gaze at a wider view: wherever I look I have before me – strictly, of course, in my mind – and 'before me' makes no sense here, since vision, for Berkeley, involves no distance whatsoever – at any rate I 'see' a constant number of minimum visibles. But how big is one and how many are there at a time or rather at a glance? Berkeley, like Locke, is very keen on the notion that we see, hear and touch precisely what we do see, hear and touch. Things not sensed, and consciously sensed (or in Locke's case consciously introspected, but that's irrelevant here), things not sensed we simply have no ideas of. That is why language, with its apparent generality, is so misleading. But then why can I not count my *minima visibilia* or give some estimate of the extent of one of them? Berkeley does, in passing, mention that the moon as seen consists of about thirty visible units, although I have not been able to confirm this. And I suppose, in his atomistic terms, one minimum visible *has* no size, since all visible size is measured in terms of numbers of these. But then if everything seen is seen and it is all so clear, give us more and more reliable numbers! How many *minima visibilia* in that green patch I judge to be a leaf? How many in the larger green and dark gray patch I judge to be part of a tree? And how many in all? Oh, but much of my mental picture is confused, says Berkeley, I can focus on only a part of it at once: that is just one of the imperfections of the visive faculty the Lord has given us. Well, then, where is all this clarity: the confidence that what is seen is seen, that the perceptible cannot be imperceptible, and so on? In particular, Berkeley insists, of course – that is the heart of his doctrine – that I have from childhood so frequently associated certain visibles with certain tangibles that I actually confuse these two very different sorts of ideas, and believe, poor benighted being that I am, that I *see* the leaves on the tree or even the valley and the hills and sky. But if what is seen is seen and what is touched is touched and there cannot be anything imperceptible in all these perceptibles, how ever have I achieved this benign confusion? How, if they are really so distinct, did I come to fuse together visible and tangible space and believe there just is space, or, better, places where things happen? Come to think of it, if it was touch that taught me distance and I then confused it with vision, how did I ever come to believe I see the sky, which I have certainly never touched? Only Chicken Little should be entitled to such a 'judgment': if one occasion – and that was all he reported – was enough to generate it. Again, Berkeley goes on and on, for instance, trying to concoct an explanation of why the moon looks bigger at the horizon than high in the sky, and his explanation stems partly from chat about the particles of air in between the eye and the horizon. Why are they not visible? And if they are not, where in the world has the '*esse* is *percipi*' doctrine got to? All this, moreover, is quite apart from the puzzle of giving to retinal images, eye muscles, motor habits and so on, an idealistic interpretation.

Or, look at another instance of the kind of perplexities to which Berkeley's 'new theory' gives rise. The doctrine of minimum visibles leads, for instance, to the corollary: "that the *minimum visibile* is exactly equal in all beings whatsoever that are endowed with the visive faculty."[3] "No exquisite formation of the eye," Berkeley declares,

> no peculiar sharpness of sight, can make it less in one creature than in another; for, it not being distinguishable into parts, nor in anywise consisting of them, it must necessarily be the same to all. For, suppose it otherwise, and that the *minimum visibile* of a mite, for instance, be less than the *minimum visibile* of a man: the latter therefore may, by detraction of some part, be made equal to the former. It doth therefore consist of parts, which is inconsistent with the notion of a *minimum visibile* or point.

Now quite apart from the question, what Berkeley might have made of the problem of the minds of mites, what can *we* make of such a piece of nonsense? Given not only the difference in size, but also the difference in the very structure of the eye, surely the mite's task of orientation, and therefore its visual world, is vastly different from ours. Malebranche described this situation better than Berkeley, but of course he was happy to admit infinitely divisible extended things, which Berkeley dare not.

Or, consider part of Berkeley's account of how we 'see' (or judge?) up and down. When you look at a man, as everybody knows, or used to think he knew, what you *really* see is an upside-down image with head on the earth and feet in the air. How *ever*, people kept asking, do you manage to think you see the man right side up? Berkeley's answer is ingenious, to say the least. We fail, he says, to distinguish the visible man and the visible earth from the tangible earth (also, one presumes, the tangible man). There is, he says, no visual up and down at all, relative to the tangible earth, since distance in any direction from the earth we stand on belongs to the domain of touch, not vision. Confused, we say the visible head on the retina is nearer the tangible earth – but, after all, if you approach the man in front of you and feel his head and feet you will find the tangible head, then tangible feet, then tangible earth – at the same time that you are *seeing* the three, head, feet, earth, in the visible realm, the other way around. But the relative visibles have simply no connection with the relation of tangibles at all! There is also more, much more, of the same sort in the *New Theory*: about the difference between the visible eye and the tangible eye (with one eye, can you believe it, looking at the picture on the other), about tangible, not visible, space as the object of geometry, and so on *ad nauseam*. But let these illustrations suffice.

For what? Only for an *ad hominem* argument against poor old Berkeley, you may say. Yet it seems to me that if you take seriously the psychological atomism on which, so far as I know, any sensation/perception doctrine rests, this is the kind of extravaganza you are bound to get into. And the moral of the story, really, is: this is supposed to be *empiricism*: the good, homey doctrine that cuts all the scholastic cackle and reverts to experience as everybody has it everyday. Did any of you ever see a minimum visible? Or touch a minimum tangible? One *could* conceive of a psychological experiment that would attempt to approximate the latter. But everyday experience this certainly is not. And of course you have all forgotten your infancy when you first learned to associate visibles and tangibles, so by now your testimony is sullied by that long association, producing Helmholtzian unconscious inferences or their equivalent, Gregorian hypotheses. Even Berkeley admits that his doctrine "will not find an easy admission into all men's understanding."[4] "However," he continues, "I should gladly be informed whether it be not true, by any one who will be at the pains to reflect a little, and apply it home to his thoughts." I have tried, as best I can – more out of sympathy,

I must admit, for Hume than Berkeley, but also because my own thinking about perception, like almost everyone's, has been infected by all those tales of upside-down flat retinal images and the inferential nature of perception which that account seems to carry with it. Yet my flat inverted retinal image is precisely what I have never seen, nor have I ever encountered a minimum visible of an aggregate of which that invisible image might be composed. The psychological atomism at the foundation of empiricism, and I think one can say of every positivistic or instrumentalist theory of science, bears, when one looks at it a moment, no relation at all to our perceptual experience as we find it when we "reflect even a little and apply it home to our thoughts."

Just look! From my hand, writing, I lift my eyes to trees, leafy valley, the hilltop, the gray clouds, framed in a narrow white window. If I look the other way, beyond obstructing furniture and the edge of the white barn wall, there are cottages, fields and a line of evergreens against the sky. Now what is most conspicuous in this sight, or these sights, is depth: what recedes from me, sitting at the table, to the horizon, more distant in one direction, nearer in the other. Depth, far from not being seen at all, is the first dimension I see in, from here to there. Moreover, *what* I see, far from being neatly distinct from the sense of touch, is ordered by its relation to where I feel myself to be. I see part of my body, for a start, and the vision prolonged from there is carried by my kinaesthetic sense of where I am, seated indoors looking out. The sight would be differently organized outdoors, in part because I would be differently located, tactually and visually, within the surrounding landscape. Merleau-Ponty seems sometimes to have a fixation with synaesthesia; perhaps he was exaggerating in order to overcome the sensationalist separation of sight and touch, much as his rhetoric about 'seer' and 'seen' may reduce to J. J. Gibson's thesis that vision always includes part of the observer's body. Be that as it may: all I want to do here, in an effort to help liberate us from the psychological atomism of the empiricist tradition, is to point out, very naively and directly, that vision, ordinary everyday vision, is organized in relation to the observer's position, in perspectival fashion, in what Gibson distinguishes, as natural perspective, from the geometrical kind. If you will just examine for yourselves any scene before you, you will find, I am confident, as Berkeley was confident you would *not* find, a multitude of ordered relations of distance, depth, obstruction, solidity, and so on. What you will find is not flat surfaces only, though perhaps a few of those, but chiefly solid objects and ordered changes in their relations: cars passing on the road, water running under the bridge, the wind blowing in the big tree. Far from being an aggregate of least bits, the visible world is constituted by a complicated network of organized objects and events, carried at the same time by the observer's bodily sense of being the one who sees. The structured richness of what we see, in other words, is its most conspicuous and persistent trait. We might contrast here with the empiricist's psychological atomism a principle of *structural pluralism*.

Even Berkeley, of course, had to link his single-sense minima somehow to one another as well as to the data of other senses. That is where his other two ruling passions, for God and for a primitive theory of signs, took over. There is language given us by the author of Nature which guides us from this to that and from here to there – if indeed such denominations can mean anything at all in a universe of spirits and ideas; in some dimensionless way, at least, they guide us, rather like the letters in an algebraic code. That sensations are sign-like has been argued recurrently; and that thesis does at least entail the correct belief that perception is not 'mere' presentation to sense, but full of meaning. A more rigorous empiricism, however, would eschew such obscurantist notions, and hold to associative connections like those advanced by Hume. For it is Hume, and Hume alone, who *really* tried to build an account of human nature out of the least felt bits to which in its British form the ideas of the New Way of Ideas had been reduced. Association is treated by Locke as a minor method of relating ideas, and for Berkeley it is transmogrified into the syntax of a language devised by God himself. But if we really take simple ideas (or impressions, to speak with Hume) as the sole matter of experience, a mechanical (or, to speak with Hume again, a quasi-chemical) association of these bits is all our mental lives can amount to. Experience, in other words, becomes thoroughly Pavlovian. Rationally, any item of experience is separable from any other; in imagination, the same disconnectibility obtains; it follows, therefore, quasi-ontologically, that the future need not resemble the past. Only blind habit forms quasi-connections among ultimately incoherent bits.

Now of course if we had only the choice – which is all Hume seems to have had, or thought he had – between God's mysterious ways on one side and on the other custom as the great guide of human life, we might well choose custom. After all, resemblances and contiguities do turn up among some items of our experience, and they do lead us to make connections that reasoning alone, in any strict sense, could not have warranted. Hume saw, at a surface level at least, the structure of prejudice and of scientific generalization: the very same structure of unwarranted extrapolation from inadequate data. A little experience yields prejudice; more of the same yields science. If we cannot have the language of God, maybe that is the best we can do.

It *would* be if we had to start with the sheer incoherence of simple impressions and ideas. If, however, we begin with being-in, our principle of connection among the aspects of experience will also look different. Again, let us make two concessions. First, many of our attitudes and beliefs may indeed be based on associative links. My mother used to take me around New Haven pointing out cats stealing from garbage cans, in order to inculcate in me a hatred of cats as dirty beasts. Although by now I like them mildly, I still shudder when my cat-loving friends allow their dear little pussies to leap upon the dining room table or kitchen counters and share their food. Whether I have any good reason for such a response to the harmless necessary cat, I have no idea. I do know I was indoctrinated. Second, at the other end of Hume's scale of non-necessities, yes, inductive inference does transcend its data; pure reason could never justify such moves from finite data to generalizations of indefinite scope. You may work up all the inductive logic you like, plain or fancy, induction still remains induction, not logic! Still, that is by no means all there is either to our everyday beliefs or to our scientific explanations. Not only association, alias blind habit, or Russellian animal inference, binds aspects of our experience together.

To see how much more diversified the ordering of our experience is, we need, once more, to start from being-in, or, as I have called it, from a comprehensive realism. And then the foundation of all the interconnections we find within parts of the world we are in will be a continuous, varied but unceasing process of orientation to the environment, both natural and human, in which we find ourselves. Orientation varies with the organism in question. In the human case, you may open your eyes, turn your head, put out a hand, get your balance, remember a rule, cry 'Slab!' – the possibilities defy enumeration. In other words, if we substitute for the principle of association what we may call *the primacy of orientation*, we will be able to note any number of kinds of connection between ourselves as culture-carried animals and aspects of our surroundings.

Watch a year-old infant learning her way around. She is exploring a definite, encompassing space with possibilities and limits: *her* space. There are things to pick up and put in one's mouth and rattle and drop, things to hold on to for walking, to climb up and creep under, to open and shut. If the space is big enough, it makes a good explorandum for quite a time. Moreover, new levels of understanding and motor competences keep emerging together. Doors are openings to a further space: why creep up and try to open specially these unless that were evident? Knowing-that and knowing-how are inextricably intertwined, as are bodily and mental powers in general. Indeed, no explorer, of any age, would have tried to find anything anywhere if his experience were really a passive flow of visual or tactile or auditory or olfactory or gustatory minima. From inside a room also, windows specially fascinate: clearly vistas to the unknown. Yet when one *is* outdoors, the scene appears to be undefined, except for a pleasant sense of motion and awareness of one's immediate neighbor. The response to human faces, moreover, differentially to parents, but also to all, is far too spontaneous to be only associative. As one becomes oneself, one is among others from the start (I'll come back to that point very briefly bye and bye). In short, there is a gradual learning one's way around *in* a house, a family, a landscape, a very different matter from the passive reception of tiny dissociative bits. As Erwin Straus insisted, the I-allon relation, the self and other interaction (where 'other', *allon*, includes objects and happenings as well as people: the organized totality of objects within which the living being moves) characterizes human sentience-times-behavior from the start. What we can substitute for Hume's gentle chemistry of association, therefore, is a pluralism of kinds of interaction. True, we may count association among these, though only as one, and relatively superficial, style of ordering. More fundamentally, however, the coherences of perception, emotion and thought as they develop together within an ordered environment dominate the blind conditionings that are all an empiricist starting point has allowed us to acknowledge. To our structural pluralism, we may add, along with the primacy of orientation, a pluralism of styles of interaction between observer and observed, or, more generally, between person and world.

Yet this is all very vague. Except for talking in a general way about orientation and exploration, I have not even hinted at what sort of processes we might substitute for association in our account of the ways experience gets structured. Nor can I, I'm afraid. I wish there were a nice little list ready to hand, like Hume's enumeration of natural relations. There are some very general pointers in the philosophical literature: in Aristotle's theory of perception, and of course in Merleau-Ponty. Since, however, we have been feeding on an inadequate, indeed, a false, psychology, it is, I believe, a better psychological foundation that we need as starting-point for our reflections. J. J. Gibson's reform of perceptual psychology, presented, for example, in his last book, *The Ecological Approach to Visual Perception,*[5] would probably provide such a starting-point. Unfortunately, I do not yet feel competent to attempt a digest of its doctrine. Indeed, until I read it a few months before writing this, I had not really understood how very radical Gibson's approach to vision was, and how completely we must abandon traditional views if we are to make sense of sense-perception.

In the light of his work, I can now understand, however, why the appeal of some philosophical writers to psychological evidence has been unsatisfactory. Both Merleau-Ponty and Michael Polanyi, for example, referred repeatedly to psychological material for support of their views about what Merleau-Ponty called the primacy of perception. Yet both lacked adequate empirical support for their sound insights. Merleau-Ponty is notorious for his over-use of a few not very good examples, and Polanyi too used to take up almost compulsively one bit of psychological lore or another as confirmation of his arguments. I was almost always somewhat uneasy about the way these came out, but had of course nothing better to substitute. First, for both Merleau-Ponty and Polanyi, there was Gestalt, an advance on associationism, certainly, but still tied, in Köhler's case at least, to a physicalistic reading. Nevertheless, it had its importance, for Polanyi in connection with the distinction between focal and subsidiary, for Merleau-Ponty in connection with the figure-background relation central to his perspectivism. Then, in Polanyi's case, there was the Ames distorted room, which fascinated him. And of course 'transactionism' sounds all very well, rather like the interactional pluralism I have just been suggesting. But the interpretation of the distorted room experiment was purely associative: we have become so much accustomed, it was said, to rooms with horizontal ceilings that in this anomalous case we see the man as small and the boy as tall in accordance with our expectation about the shape of the room. (Haven't we generally seen men larger than boys as well, one wonders?) For years, also, Polanyi kept harping on the Innsbruck school of Ivo Köhler and inverting spectacles. There was allegedly one member of the group who resisted the standard Berkeleyan interpretation of their work, and that was supposed to be especially revealing. But even this heresy never seemed quite right. And finally there was something called 'subception,' which I gather had better not be mentioned.

In addition to the scarcity, and now the novelty, of good psychological material to support a new perspective on perception, moreover, there is the general problem which I at least find puzzling, of how it is that empirical work in any area bears on philosophical issues. I do believe there is a philosophical problem about perception and the relation of perception to knowledge, which experimental psychologists, however gifted and however original, cannot altogether solve for us. Philosophical problems are metaproblems and cannot be turned over to experimentalists for their solution. Yet neither do our problems lack all connection with everyday experience or with the fruits of scientific inquiry. In this case, at any rate, where we have, as I have said, been deceived by a mistaken psychology, we could certainly use some information about a more adequate doctrine to help us rethink our philosophical questions. At this point, however, I can only make this bald pronouncement and conclude that, whatever positive concepts we use to conquer Hume's associative scheme, with its operationalist or behaviorist consequences, we can do this only if we have abandoned firmly, once for all, the doctrine of least psychological bits – and that not in favor of an equally antiquated physicalism, but of a fresh, ecological approach. Only then can we begin to build a more adequate account of our dealings with the world.

Further, this holds for all such dealings, both everyday and esoteric. For it is only if we can make sense of sense-perception that, in turn, we can make sense of the cognitive interconnections built on its foundation. Although knowledge, including scientific knowledge, elaborates sense perception and embeds it in complex theoretical structures, it necessarily returns to perception as its base and is even, in the last analysis, an extension of it. I say this here chiefly to link my remarks about perception officially to my official subject: empiricism and the philosophy of science. Confirmation of this thesis and its expansion in more substantive form can of course be found in a number of places. Apart from my principal source, the work of Merleau-Ponty, I may refer you for example to a recent *Monist* paper by Harold Brown on 'Observation and Objectivity,' as well as to his book.[6]

Even if I defer to psychology for the moment, however (though not ultimately, if you please, in the spirit of a naturalized epistemology) – even if I defer to psychology for the moment, there are nevertheless several additional philosophical points I should like to make briefly, in conclusion. I have mentioned all of them in passing, but they deserve some special notice.

First, if the follies of empiricism, and hence of philosophy of science, have derived, with a twist, from the Cartesian separation of thinking mind and inert matter, they have exhibited in particular a special error consequent on the Cartesian reform, and that in paradoxical fashion: I mean, the sharp separation of action and passion. Reid congratulated Descartes especially on this division: mind is wholly active, matter wholly passive, what a wonderful insight! Yet mental content, on the other hand, in Locke's supermarket model, becomes wholly passive, too, as everything non-mental, namely matter, was now generally supposed to be. And when Berkeley's 'notions' fade, as, shadowy creatures from the start, they easily do, the mental itself goes passive. True, nowadays there are all those clever stunt riders chasing one another around the action theory corral; but so far as I can tell they have little effect on thinking outside their special circle. And besides, they still belong firmly with the Cartesian alternative: Newtonian nature and the choosing will, nearly Sartrean in-itself and Sartrean for-itself, are the pair they appear habitually to deal with, so that when reasons are not pure reasons they lapse into causes, and so on. All that, I am happy to say, is beyond the scope of my present remarks. All I want to note here is that, as far as mind goes, it is the radical separation of action and passion, whether in favor of will as agency on the active or sensation/perception on the passive side, that has been misleading for the interpretation of any cognitive process. All our experience, including scientific research and scientific discovery, exhibits both active and passive features from the very start.

Incidentally, it is paradoxical, too, that where the empiricist mind turns passive, like the bare inert matter of *res extensa*, the would-be objectivism of this tradition, the attempt to run down reason and seek the real world out there where science finds it: that effort itself succumbs to subjectivism: the real world itself reduces to sensed bits. But that is by the way.

Another point that I have alluded to in passing is the impotence of empiricism to take account of the social character of science. As I put it earlier, ghosts do not communicate and neither do machines, however cunningly we may use them for communicative purposes. But science is a family of practices – 'family' in the Wittgensteinian and 'practices,' I hope, in the MacIntyrean sense.[7] And practices are by their nature rooted in community and permit communication only on that ground. Each science requires from its initiates a long and arduous apprenticeship, in which they master the language of their discipline, its leading interests, beliefs and puzzles, its techniques of model-making and of calculation and, in most sciences, of experimental design – all these factors being assimilated by the novices in intimate interaction with one another as well as with their teachers. Thus it is only a philosophical approach which accepts the sociality of human nature from the very beginning that can begin to deal adequately with the cognitive claims of science, let alone its relation to other commitments of the culture to which it belongs. For each practice has its own morality, which in turn must be accredited by the larger ethic of the society in which it has its life.

Finally, psychological atomism, in Berkeley and Hume at least, entailed nominalism, or in Locke a theory of general ideas so fatuous as to draw nominalism inevitably in its wake. So, in philosophy of science, the model of observed bits, laws, prediction and again observation was allegedly a model comprised, on the ground floor, of particulars only and allowing non-particulars solely by convention or linguistic fiat. From the seventeenth century on, universals had been decried as fictions of the schools, and finally with the world pulverized to particulars we had those headaches about the solubility of salt, brown ravens and what not. Yet the alleged particulars of empiricism were not really *just* particulars – how could they be? Locke, as I remarked earlier, quite blatantly failed to distinguish particulars from universals. He talks as if the ideas of yellow, hard, fusible, soluble in aqua regia, were all particulars, and only the idea of gold general. But it is precisely predicates that are general: that ring on Locke's finger was a single entity, all right, yet he could not have recognized it as such without recognizing its predicates, yellow, hard, circular and so on, which are general. The same goes for Berkeleyan ideas and Humean impressions: if each such particular were *only* a particular, God's sign language could not function, nor could the gentle force of association operate, since the dissociation of each sense impression or idea would be complete. Surely the same impression, or idea, never returns, only its like or its neighbors, which must be recognized as of the same sort again. I have no intention of embarking herewith on a treatise about universals – for one thing, I cannot think of any new arguments, there being plenty of valid ones about for those who want them, and those who do not will not listen anyway. I want only to emphasize the would-be nominalism of the empiricist tradition as one of the major reasons for its failure to interpret adequately knowledge in general and scientific knowledge in particular. Were we not in a world where, as Peirce put it, some generals are real, we would have neither language, including the languages of science, nor action, including the practices of science, nor any knowledge of anything at all. But we do know quite a lot, one way and another; what we need to learn, however, before we can know better what knowing amounts to, is to rid ourselves, once and for all, of the nominalist bias of empiricism, as well as of its other prejudices, all of which still support the tangled web of pseudo-problems characteristic of most philosophical practice, not quite so thoroughly by now in philosophy of science as in philosophy, but still pervasively enough.

It is to that difficult learning process that I wanted, through this backward look at our empiricist ancestors, to give some small encouragement. Granted, not every one likely to read this essay needs the foregoing tirade, yet Cartesian-empiricist thought with its entangling consequences still exercises a mighty influence in philosophy and at least by spin-off in philosophical debates on the methodologies and claims of science. So even if I cannot singlehandedly decapitate that Gorgon, I hope I have at least shown what kind of mirror might help some more competent Perseus to aim deftly at the monster's throat and free us once for all from her petrifying stare.

NOTES

[1] Thomas Reid, *Essay on the Intellectual Powers of Man*, ed. B. Brody (M.I.T. Press, Cambridge, 1969), pp., 138–9.

[2] Richard Rorty, 'Incorrigibility as the Mark of the Mental,' *Journal of Philosophy* 67 (1970), 399–422.

[3] George Berkeley, *A New Theory of Vision*, section 80.

[4] *Ibid.*, section 112.

[5] James J. Gibson, *The Ecological Approach to Visual Perception* (Houghton Mifflin, Boston, 1979).

[6] Harold I. Brown, 'Observation and the Foundations of Objectivity', *Monist* 62 (1970), 470–481.

Harold I. Brown, *Perception, Theory and Commitment* (University of Chicago Press, Chicago, 1979).

[7] A. C. MacIntyre, *After Virtue* (University of Nortre Dame Press, South Bend, 1981).

I. C. JARVIE

REALISM AND THE SUPPOSED POVERTY OF SOCIOLOGICAL THEORIES[1]

For all the Idols of the Mind or Profession regnant today the worst is that which Bacon might have placed among his Idols of the Theatre: the belief, first, that there really is something properly called theory in sociology, and second, that the aim of all sociological research should be that of adding to or advancing theory. It is a truth we should never tire of repeating that no genuinely good and seminal work in the history of sociology was written or conceived as a means of advancing theory – grand or small. Each has been written in response to a single, compelling intellectual problem or challenge provided by the immediate intellectual environment. William James did not err in labelling as "tender minded" all systems-builders, whether religious or law, and placing under "tough minded" those who welcome and deal with life in its actual concreteness.

Robert Nisbet, *Sociology as an Art Form*, New York: O. U. P. 1976, p. 20

My title may be misleading. This paper will argue that contemporary sociological theories are not impoverished. On the contrary, it is my observation that contemporary sociological theories are rich and diverse.[2] There are abroad Marxists, functionalists, structuralists, phenomenologists, symbolic interactionists, ethnomethodologists, conflict theorists, labelling theorists, critical theorists, and so on. That they battle and proliferate strikes this philosopher of the social sciences as healthy, fruitful and exciting.[3] As an editor, I never know what is going to flop into my in-tray next. In addition, there are certain maverick figures who are doing incredibly illuminating thinking at the theoretical level, especially Edward Shils, Raymond Aron, Ernest Gellner and Erving Goffman.[4] So, much of my space will be given over to explaining how such richness can be denigrated and the claim of theoretical poverty made.

Poverty, like health or wealth, is a comparative concept, a question of relative deprivation.[5] If the state of theoretical sociology appears to some to be impoverished, it must be being measured against certain expectations; in trying to show that it is anything but impoverished I shall need to expose and criticise those expectations and try to replace them with more defensible ones.

Expectations of sociology have run pretty high. Auguste Comte, our revered High Priest, placed sociology at the top of the hierarchy of the sciences; the king, as it were, to mathematics, which is commonly called 'the queen of the sciences.'[6] As befits the top science, it has been expected

R. S. Cohen and M. W. Wartofsky (eds.), Epistemology, Methodology and the Social Sciences, 107–121.

that sociology would explain and predict human action with a precision and depth comparable to theories in the natural sciences, especially physics. It is significant that less mathematical, less abstract and less precise natural sciences such as geology and biology were not taken as exemplary. Indeed, such is the prestige of physics that biology and geology themselves labour under its shadow. By the standards of physics we are all weighed in the balance and found wanting.

There are at least two very good physical reasons why it is inappropriate for other sciences to take physics as an exemplar. The first is *abstraction*. Physics is the most general natural science, its domain is the entire range of phenomena in the natural world. Everything obeys the laws of physics. Because of this scope, physics operates on a very general and abstract level. Or, rather, we describe it this way because, as physical systems, we human beings are of middling size. The very large galactic systems and the very small molecular and atomic systems are literally invisible to us unaided, yet we believe that many of what physicists call the most fundamental of physical processes take place in those ranges. The physics of those phenomena appear somewhat remote from our own, quotidian level.

The second reason why physics should not automatically be taken as an exemplar for other sciences is *isolation*. When physics is admired for the precision and depth of its theories, what is usually being thought of is not its treatment of heat and light, or sound and electricity; it is, I suggest, mechanics in general and celestial mechanics in particular. Nuclear physics is too arcane for most of us; whereas predicting sunsets, sunrises, comets and eclipses is both spectacular and accessible. Unfortunately, even within physics, there is something unique about celestial mechanics. For purposes of calculation, the solar system can be treated as a physically isolated system, i.e. one on which the effect of outside forces is negligible. Such a system is wonderful to work with but, to say the least, not very common in the known world. In particular, no such isolated systems exist in the social world. We may, for heuristic purposes, attempt to treat social systems in isolation to see if that bears any fruit. The most successful example of this clearly being the case of economics. While that discipline is impressively elegant, precise and algebraic, there is much debate about the testability and applicability of its results.[7]

From these arguments I wish to draw the following conclusion: we should guard against excessive expectations of the social sciences. To expand: we cannot expect the abstraction, precision and hence mathematicisation so familiar in physics; we also cannot expect the depth, in the sense of the revelation of unsuspected hidden processes that totally govern us; and, finally, we cannot expect anything like the degree of consensus that is to be found in the natural sciences. There is quarrel, party and confusion there too, but there is bound to be more in the social sciences. For one thing, every man cannot but operate with mental models of the social world and hence is a social scientist and knows himself to be such; every man cannot but operate with mental models of the physical world and hence is a physicist also, but does not know himself as such. Hence we might expect greater diversity of opinion and more vigorous debate in the social than in the natural sciences. Another reason for this might be the very direct connection between views on how social life works and views on how social life should be. There is no problem of how nature should be, in physics; nature is as nature is. It is different with society. People are as interested in how things should be in society as in how they are; perhaps more interested.[8]

So, while the social sciences may not be able to be so abstract, precise, surprising and convergent as the natural sciences, there is every reason to expect they will be theoretically rich and vigorously debated.

Let us not go too far in lowering our scientific expectations. Although social systems are neither as general nor as isolated as the solar system, and indeed despite their being among the more complex systems known in the natural world, and despite their complex patterns of interaction with each other and with the so-called environment, they nevertheless have order and coherence. Remarkably, human beings are highly predictable, as demographers and actuaries will attest. The shock of recognition when we read the brilliant – albeit atheoretical – observations of Erving Goffman, or Harold Garfinkel and his students, derives from this predictability. Every time we go to the bus-stop or the airport without phoning ahead we display our confidence that most of the time our fellow humans are predictable, and hence that man-machine systems are predictable also.

How can this be so? How can the uniquely complex human being, strongly interacting with his environment, nevertheless be highly predictable? Clearly, there is *some* means of coordination, some means of simplifying and regulating interaction. In my view the means of coordination is the capacity of the human mind to model all the elements of its environment, not excluding itself. We build, I argue, models of how the world seems to us, models we then use to guide us in acting to achieve our aims.[9] Indeed aims themselves are models of a sort, since they involve envisaging or modelling a future state of affairs. But if millions of heterogeneous human beings were busily modelling their environment there is not the slightest *a priori* reason to suppose that they would come up with anything other than a bewildering chaos of variations. What coordinates our model-building efforts? Two obvious factors: structural similarities in human beings connected with their very capacity to model themselves;[10] and reality. The notion of reality, nature and its objects given, as they are, independent of the perceptions of men, is not just a metaphysical postulate: it is an essential ingredient of the success of life in society. Ontological realism is not an assumption we can blithely dispense with.[11] It undergirds our essential coordination and the fundamental rationality of our model-building. Supposing, as we do, that our models are models of something, namely a unitary and relatively unchanging world, we act to modify and improve them, to, as we say, 'get it right.' Our inclination to do this stems from the very obvious point that if we act on mistaken information, i.e. incorrect models, we are unlikely to achieve our aims. If, in coming to deliver this paper, I was misinformed about the date, time or venue, its accomplishment would be threatened. To say to my hosts that 'sociology is not concerned with a "pre-given" universe of objects, but with one which is constituted or produced by the active doings of subjects'[12] would be a poor excuse. Indeed, we might reflect on the very fact that accomplishment of my talk is facilitated by the calibration of time and space into days, months, hours, streets, buildings and rooms, yet that calibration is clearly a social construction, a set of conventions widely accepted around the world. The conventionality of our calibration does not infect its objects: and yet, time and space, and the entities that can be plotted on those coordinates, are to a certain extent pre-given. Languages, those other widely diffused conventional systems, are even more remarkable, since they not only accomplish the task of coordination, but they even permit discussion of the possibility that there is now a world to coordinate our discussions of and actions in.[13]

There will be those who contend that what I have described as reality, what is really there, to speak naively, is simply the general consensus of opinion. This looks tempting. However, it confuses a test of reality, i.e. what most people think it is, with reality itself, i.e. what it in fact is. It is obvious that if what were actually the case were the same as what a majority thought was the case, then there could never be any rational change of views, especially of the kind we call correcting. Yet this happens all the time: the majority has to revise its opinion because of the arguments often enough of a single hold-out against orthodoxy. In the end, what is persuasive, what is the strongest argument there is, is that the hold-out is correct about the world.

What I am saying is that our capacity to explain and predict what is happening in society, to do sociology, can be traced back to our model-building activity, to our all being ur-sociologists. We need to be sociologists of no little skill to function effectively in society. Goffman has shown how even the mad must build a coherent and model-able social system within the 'total institution.'[14] The fundamental problem of and for the psychotic is that his model of himself and also of his surroundings is so discrepant that his behaviour is inexplicable and unpredictable on ordinary modelling assumptions. To pretend that he is possessed of deep insight is to elevate honest disagreement into bad philosophy. The madman has succeeded in being classified as such because the models on which he acts are at variance not just with the way others model, but also with the way the world truly is.[15] A less fanciful model will make it easier for the madman to accomplish his aims, and less liable to classification as someone irresponsible.

Madness is an example of a social phenomenon that challenges the common-sense sociological models of everyday folk. Acting on those models we would reassure the paranoiac that not everyone is against him, attempt to cheer up the depressive with some bromide. But these moves fail, can even make things worse. We had to wait until Freud came along and began the work of building a whole sub-field of sociology called psychology before we could reconstruct the modelling that might be going on. Freud showed us how mad behaviour becomes explicable and hence predictable if you infer that the person has and therefore inhabits a different model of the world than the rest of us. In particular, the models of persons and the structure of the nuclear family and of the self are often what is most seriously warped, albeit in an effort at self-defense. Equipped with this insight we can de-mystify madness.

Much the same happened in the creation of anthropology. Parochial or ethnocentric models of other people's social behaviour, including speech, yielded little of explanatory and predictive power when applied to the exoticism of preliterate tribesmen. Indeed, in early contact situations aboriginals were described as living in a confused, chaotic, and savage manner, possibly incapable of coherent speech. This is a wonderful example of projecting the problems and bafflement of the investigator onto his subjects. Again, new and much more sophisticated models had to be built, incorporating the modelling assumptions of the aboriginals. These models constitute the corpus of theory of social and cultural anthropology.[16]

Were I out to be really provocative, I could suggest that economics is the name given to the model-set that treats the wealth system to the extent that it can be analytically separated from the other systems in society, and that political science is the name for the somewhat less successful set of models that attempt to isolate the power system. Each is a branch of sociology.

Turning aside from such reckless attempts to be provocative, let me turn back to the idea that the explicability and predictability of social life stems from our innate capacity to be sociologists, i.e. to model ourselves, others and the system in which we all interact. This would give rise, one might naturally expect, to a great wealth of sociological theorising at the popular and at the professional levels. Whence then the charge that there is a poverty of sociological theorising? The cry comes partly from uninformed outsiders who just don't know enough to be taken seriously; and partly from sociologists and their kin. It is the latter I want to diagnose. There are those locked into rigid theories from which they cannot be budged; there are those so elusive no one can tell whether they operate with theories or not, still less what those theories might be; and there are still others who are explicitly atheoretical and/or antitheoretical. All three categories can, I believe, be explained together: they are reactions to what is seen as the failure of sociological theory. As I have already argued, such reactions are overreactions based on excessive expectations. Towards the end of this paper I shall try to show how more modest expectations, combined with an acceptance of refutation as a learning experience, can defuse these over-reactions. In the meantime, I postulate that all three types of theoretical inadequacy stem from disappointment.

Characteristically, when a cherished view is decisively refuted, four different kinds of reaction can be observed. There is the reaction of stubbornly holding on to the cherished view, this I shall call dogmatism. There is the reaction of modifying the cherished view towards vagueness and hence irrefutability, this I shall call mysticism. There is the reaction of despair: if this cherished view can succumb then all views must succumb, so what is the use; this I shall call *scepticism*. And finally there is the uncommon reaction of saying that the refutation at least has taught us something, and this I call *critical rationalism*.

Work on millennial religious cults provides crisp illustration of these reactions. Failure of the end of the world to arrive on schedule has the following effects. Some cultists drop out and become sceptical of this religion or all religions; some cultists go right on believing as before, perhaps slightly altering the prophecy; some seek hidden meaning in the messages and prophecies; and almost none simply conclude that, well, now we know that the world is not going to end.[17] A remarkably similar spectrum of reactions could be observed in the United States as the Watergate scandal unfolded. What was disclosed was a challenge to believers in Richard Nixon and to believers in the integrity of the office of President. The rational reaction that Richard Nixon was shown once more to be a scoundrel and that nothing magical in the office of Chief of State prevented its seizure and misuse by unsuitable persons, was not only relatively rare in the welter of dogmatism, mysticism and scepticism (the latter sometimes diguised as cynicism), but was in some quarters actually ridiculed as naive.

My suggestion is that sociologists have been disappointed with the performance of sociology and that their disappointment is a function of their excessive expectations. That we have had the good fortune in the past to model society with some little success should not delude us into thinking that we know it all. Then, if we do come a cropper, we need not dogmatically pretend nothing happened; mystically proclaim that events surpass ordinary understanding; or sceptically declare that the quest to understand is hopeless.

Instead, we can try to learn from experience. The rise of sociology in the modern world can be seen as a case of society learning from the hard knocks of experience. Learning from experience consists in learning that something we thought we knew, some mental model of the world that hitherto served us well, is false, i.e. no longer explains what is happening. Our policy in the face of this discovery obviously should be to find out how extensive the failure is and to see whether a new model can be developed to fit these new cases and the old ones, or whether models will have to be proliferated, or what not.

Before the rise of modern science, I would maintain, all societies operated under comprehensive world-views that explained and predicted what happened in society and also, conveniently, explained failures to explain and predict. A new situation was created with the rise of modern science and its world-view, or should I perhaps say, its opposition to comprehensive, all-explaining world views. Science is less a world-view on the same level as its magical and religious predecessors, and more a meta-level doctrine about how we should proceed, namely, by casting models, theories or world-views, and then tempering them on the anvil of recalcitrant experience.

The rise of sociology is a specialisation of this general process relative to the modelling of society. The triumph of the mechanical world-view of science and its embodiment in industrialisation and industrial society presents a fundamental challenge to all traditional models of society. No longer do official religions or hierarchies sufficiently assist displaced industrial man to model society. The rise of sociology is the rise of a group of specialists to help in the formulation of social models that can cope with change and learning from experience in the same way that natural science has. Not all sociologists will accept this description. I would nevertheless maintain it to be a correct account of one function of the institutuion of which they are part. The folk sociology of ordinary people having proved inadequate, we must beware of allowing the folk sociology of sociologists to replace it. Folk sociology is that received from the tradition and trying to preserve it. Culture clash, social change, industrialisation, catastrophe should neither be simply ignored, nor glibly reconciled with tradition. The life of man has taken a decisive new turn in the last three hundred years, and sociology's job is to try to understand it.

Despite the obviousness of this commission, sociologists are prone to over-optimism and hence to over-reaction to failure. The fundamental failure is not to have been able to perfect society. The optimism engendered by the social and intellectual liberation that began in Europe roughly in the fifteenth, sixteenth and seventeenth centuries was enormous. The optimism for the future that underlies the American and French revolutions and much social and philosophical thought in the eighteenth and nineteenth centuries is there for all to see. Yet many sociologists seem disappointed; neither society nor sociology has turned out as over-optimistic expectations allowed.

Let me flesh this out a little by looking briefly at four founding fathers of sociology, their problems, their aims and their eventual failures, and let me argue that it is from these intellectual failures that we learn, and hence that the name 'failures' may be a misnomer.

Alexis de Tocqueville was intrigued by the problem of how this new phenomenon, democratic society, could maintain social order without the institutions and rules of aristocratic society. Karl Marx struggled to understand how Adam Smith's arguments that minimally regulated capitalism would result in the greatest possible wealth for nations, failed to predict the exploitative poverty and degradation of large masses of people. Max Weber tried to show that there was a connection between religion and the rise of capitalism and hence that Marx's materialism was false; and he also argued that order in modern society rested on a new kind of legal-technical rationality, embodied in industry and bureaucracy. Emile Durkheim was puzzled by social regularities, by the constancy of suicide rates and by man's capacity to classify.

One would not call these problems of these great sociologists negligible

– a science with four Galileos hardly needs to doff its hat to its sisters. Yet the charge of theoretical poverty is still laid; sociology does not come up to scratch. What, however, is scratch? My suggestion is that scratch has been thought of as perfect understanding leading to perfection of society.[18] Even as I write this down it seems absurd. But I know of no other explanation. Because sociology does not yet have the truth, and because we need the truth to set us free, sociology is decried even by its own. Marx, although explicitly denouncing utopianism, nevertheless vaguely hinted at society coming to a kind of perfect resolution as sociological knowledge (he called it 'class consciousness') became widely diffused and led to appropriate action. For this reason, Marx's intellectual influence has been somewhat baleful.

When a social scientist offers to try to solve a problem, let us say unemployment, it is sloppy logic and bad faith to treat him like the Pied Piper and, when he has finished, say, 'yes, but what about the inflation?' An inventory of the social and intellectual problems that have been eliminated, reduced, minimized in industrial society would be an extensive one. The success of sociological reflection should, I would argue, be measured against that, not against what remains to be done.

Yet, it is my conjecture that it is precisely against the inventory of what remains to be done that sociology is tested and found wanting by those who declare it to be theoretically impoverished. Democratic society may have ensured that the privilege, snobbery, inequality and lack of social mobility of aristocratic society were overcome. To now charge that democracy is not without its faults, that some of these things have re-emerged in new form, together with imperialism, militarism, racialism, etc., is to hark after an unreasonable perfection. It is also to forget that Tocqueville foresaw some of it. Marx and Methodism, it is said in England, produced the welfare state and all the hopes that attached to it. Alas, those hopes were dashed, for the welfare state not only had social problems, but it even has some that survive from the previous state of affairs. Weber's ideas of the connection between religion and economic organization, and on the necessary growth of the bureaucratic mode of organization continue to hold up. To castigate Durkheim's explanation of order and cohesion by pointing to the conformity and privatisation of affluent society is to mistake new or modified problems for old, to fail to give credit to theoretical sociological thinking where it is due.

The logical name for the fallacy we see at work here is shifting your ground. It is one of the most pervasive fallacies and also one of the most difficult to detect. Why does it happen? Why should there be a denigration of sociological thought when the achievements are so considerable? The answer by now should be fairly obvious: dissatisfaction stems precisely from success. In sociology we call it the revolution of rising expectations; that is, it is when conditions are getting better that revolutions come about. It seems as though in hopelessly oppressed conditions discontent is contained, inarticulate. Only when a small improvement raises hope for further improvement, and when that improvement is felt to be too slow, only then does discontent and rebellion break through. How ironical then that sociologists implement their own theory: they participate in the process of getting some improvements in society and this makes them find the remaining unrelieved conditions intolerable. Under the influence of sociologists and their cohorts in the other social sciences our society has fought the war to end all wars, the war on poverty, for the establishment of the right to work, for the four freedoms, for desegregation, equitable immigration, and so on. Alas, a small lesson from epistemology might help here: the successful solution to a problem always leads to the appearance of new problems and a revision of the schedule of urgency with regard to still unsolved ones.

Indeed, problems breed like nuclear reactions. When P is solved by S, S itself will have components (entities and relationships) $E_1 \ldots E_n$ and $R_1 \ldots R_n$: and each of these components can in its turn yield a new problem-cluster. If I explain the splitting of the atom by a model that says an atom is not an indivisible basic entity, but in fact a complex made up of three entities, protons, electrons and neutrons, the problems now arise as to what the exact details are of their spatial arrangement, their energy and other characteristics, what kinds of forces hold them in place and so on. The whole process reminds one of medicine. A disease like diphtheria ravages children. An antidote is found, but then some react badly to the antidote, whether it be immunising serum, drug therapy or hospital treatment. Next thing you know there are hospital and drug-induced diseases, etc. Only by keeping the original aims in mind is it possible to avoid scepticism in the face of this failure of disease to disappear, physics to reach a resting place, society to be free of problems.

Given then that sociologists may have been disappointed by their own subject, we can now analyse their reactions in greater detail utilising the three characteristic reactions to refutation discussed earlier. All three view a refutation as a failure, a defeat. As I shall try to show, only by reversing this and seeing refutation as a success, as a clear growth in our knowledge, can these reactions be avoided.

Reaction one: *dogmatism*. The despair of dogmatism lurks in some of the most radical sociological theories, such as Marxism and behaviourism. Despite repeated refutation of these theories in detail, they still have many stubborn adherents. This is helped by the fact that both are highly programmatic theories, that is, theories that promise that a great deal more will be forthcoming. What has not yet been forthcoming can hardly be refuted. It is also part of the appeal of these theories that they embody a total world-view and world-views do not, indeed cannot, get straightforwardly refuted.[19]

Reaction two: *mysticism*. We are familiar with the mystical reaction in religion, especially when internal contradictions are exposed and devotees seek refuge in the ineffable. The most popular language in which to express mysticism in the social sciences is Hegelian. Hegel has been described even by his own followers as scarcely intelligible, but a lot of it is quite impressive-sounding. We find mystical Hegelian talk among some Marxist social scientists, many phenomenological social scientists, among critical theorists, and in the group influenced by Alan Blum. Some critics contend that nothing is being said; others that banalities are disguised in high-sounding language; still others that there are unplumbable depths of confusion involved. My explanation, that they are reacting to the disappointment of refutation by making their doctrine immune to it, strikes me as the best reconstruction.

Reaction three: *Scepticism*. Scepticism about the very possibility of sociological theory is far and away the most common of the three reactions today. As representatives of this scepticism I shall instance analytic philosophy, atheoretical empiricism, ethnomethodology and the sociology of knowledge. Let me expatiate a little on each.

Analytical philosophy took a sociological turn with the advent of Wittgenstein. Although, so far as we know, he knew no social science whatsoever, he did try to ground the study of concepts and categories into the matrix of social life to which they belong. It was from reflection on how language is used to mean that he came to his famous conclusion that the meaning is the use and the use involves studying the forms of life in which the language is embedded. But then Wittgenstein's disciples neatly reversed the equation and argued that studying social life is studying meaning, studying meaning is studying concepts and rules, and that studying concepts and rules is an *a priori* activity reserved to philosophers.[20] A science of society is not possible because the events in society are meaningful in a way that events in brute nature are not. Quantification, precision, explanation, prediction, all these become category mistakes. With minor variations this line of argument has been pushed by Peter Winch, A. R. Louch and Keith Dixon.[21]

Atheoretical empiricism is a less articulate position, as it denies that any convincing rationale is possible for collecting data. Just doing the survey, tabulating the results are difficult to justify with the bald assertion that the author finds them interesting. And the inductivist defense that only when we have enough data classified and tabulated will we be ready to generalise, i.e. theorise, has worn a little thin, even without the by now well-known existence of knock-down arguments against it.[22]

Ethnomethodology, in its curious way, also eshews theory for observation: it adopts the premise that society is constituted by those in it, and that their rules and practices are what sustain the social order. Whereas I have contended that people are theoretical sociologists, ethnomethodology emphasises the indisputable fact that they are practical sociologists, too. There is also a convergence with analytic philosophy in that both give pride of place to concepts and rules employed by the social actors (or members), rather than the artificial and imposed categories of the sociologist.[23]

Finally, the extreme and articulate scepticism of the sociology of knowledge as developed by Berger and Luckmann and some of their even more radical followers. Their radicalism consists not in delimiting the possibilities for theorizing in the social sciences, as analytic philosophers, atheoretical empiricists and ethnomethodologists do, but rather in arguing that sociology shows that a theoretical science of society is impossible because a theoretical science of anything is impossible. The so-called real world itself is held to be a social construction and the activity of scientific investigation of it is merely a disguised projection of sociological necessity. The world seems the way it does to us because of the conceptual spectacles we take from our social conditoning. Any science with pretensions to depict the world as it really is, is naive. For a subject to connive in this way in its own self-destruction is a radical scepticism indeed.[24]

There is a fourth reaction to refutation which diffuses the disappointment and which I recommend, namely, allow refutation to teach you: i.e. learn from experience. If a favourite theory flounders on a nasty, recalcitrant fact, comfort yourself with the thought that that step backward for you may be a giant step forward for mankind. But I am getting too sententious. Such a policy of critical rationality is psychologically hard to adhere to, and yet is characteristic of what we call the scientific attitude. It is a pity that it is not more often consciously given a chance. Social theory is constantly subject to the test of being implemented by politicians in programmes of social reform. Because politicians embrace ideologies, and because for some unknown reason they never want to admit to making mistakes, the results of these tests are rarely discussed critically and rationally. Reform is almost never implemented with a clear understanding that if such and such is achieved that will be success, if it is not achieved, that will be failure.[25]

Although the attitude to sociological theorising is rarely what it could be, I nevertheless want to declare the state of the art rather healthy. My point would be that although poor philosophies and methodologies can perhaps inhibit the growth of knowledge, the logical force of the problems and the facts themselves seems to be enough to prevent our endeavours grinding to a halt. This cannot be allowed to conduce us into a complacent tolerance of poor philosophies. An especially poor philosophy I wish to chastise is what might be called the consensus model of science; it is opposed to the dispute or debate model that I espouse. Talcott Parsons, who died recently, was a man who did much for sociology, but who also did it great harm with his fantastic thesis that all previous sociologies could be lined up in kind of convergence towards agreed scientific findings.[26] Another and perhaps even more baneful convergence model has hung over us since 1962, when Thomas Kuhn published his magnum opus.[27] In that work we are told that the squabbles and debates in the social sciences are precisely what mark them off from (and as inferior to) the natural sciences. On the awesome authority of his expertise as an historian of celestial mechanics, Kuhn claims that real natural scientists do not quarrel about fundamentals, do not argue and debate, but quietly get on with the business-as-usual of puzzle-solving. Kuhn is a good sociological observer, and has impeccable scientific credentials.[28] Yet his view should be contested. Some bemused sociologists have happily convinced themselves that our competing theoretical systems and schools of thought are really paradigms, under the auspices of which puzzle-solving can be carried out.[29] This will not go through: Kuhn thinks multi-paradigm situations are rare, and are a sign a discipline hasn't settled down yet to scientific work. Better to consider the possibility that sociology refutes Kuhn, making one re-examine his claims about sciences as authoritarian in structure and routinised in activity. When we do this, examples are to be found everywhere of quite fundamental debate, and plenty of scientific schools where dissent not consensus is encouraged.[30] Perhaps the philosophy of the social sciences takes its most urgent problems less from the struggles of living science and more from the need to defend such honest toil from the bad advice of philosophical amateurs. In doing so, the philosopher can offer philosophical, not scientific, advice. Any science concerned about theoretical poverty could best initiate a debate about it – this is the rational policy – despair is never the rational option.

York University, Toronto

NOTES

[1] This paper was written while I was on sabbatical leave from York University and an Associate of the Center for Humanities, University of Southern California, in receipt of research monies from the Canadian Humanities and Social Science Research Council (no. 451–790154). I am grateful to those institutions. It was read in the lecture series 'Philosophy and Sociology: Confrontation and Rapprochement', University of Dayton, October 9, 1979, and to the Department of Sociology Colloquium, UCLA, November 7, 1979.

[2] See John C. McKinney and Edward A. Tiryakian (eds.), *Theoretical Sociology, Perspectives and Developments* (Appleton Century Crofts, New York, 1970).

[3] I must qualify this generalization by reference to the work of one of my students, Jean E. Saindon. In his Ph. D. dissertation he reports that there is much smoke and very little fire in the running debates between empiricist and idealist (or interpretative) sociology. See his 'Epistemological Dogma in Sociological Thought', Ph. D. Dissertation (York University, Toronto, 1979).

[4] Edward Shils, *The Intellectuals and the Powers* (University of Chicago Press, Chicago, 1972); and *Center and Periphery* (University of Chicago Press, Chicago, 1975). Raymond Aron, *Main Currents in Sociological Thought* (Basic Books, New York, 1965); *Eighteen Lectures on Industrial Society* (Weidenfeld and Nicholson, London, 1967). Ernest Gellner, *Thought and Change* (University of Chicago Press, Chicago, 1964); *Legitimation of Belief* (Cambridge University Press, Cambridge, 1975); *Spectacles and Predicaments* (Cambridge University Press, Cambridge, 1979); *Muslim Society* (Cambridge University Press, Cambridge, 1981);Erving Goffman, *Frame Analysis* (Harvard University Press, Cambridge, Mass., 1974).

[5] W. G. Runciman, *Relative Deprivation and Social Justice* (Routledge and Kegan Paul, London, 1966).

[6] The dynastic marriage has not been consummated in sociology as it has, e.g. in economics and geography. That should not be a matter of concern, as I shall argue below.

[7] G. C. Archibald, 'Method and Appraisal in Economics', *Philosophy of the Social Sciences* 9 (1979), 305–316.

[8] See my 'Nationalism and The Social Sciences', *Canadian Journal of Sociology* 1 (1976), 515–528.

[9] This thesis is argued in detail in my *Concepts and Society* (Routledge and Kegan Paul, London, 1972).

[10] I do not need a Chomskian innate capacity to learn language (as per his *Cartesian Linguistics* (Harper and Row, New York, 1966), all I need is a disposition to survive in the environment and hence to 'develop' tools that aid that quest.

[11] See my 'Cultural Relativism Again', *Philosophy of the Social Sciences* 5 (1975), 343–353.
[12] This is Rule A, One, of Anthony Giddens' *New Rules of Sociological Method*, (Basic Books New York, 1976), p. 160. The original is in italics.
[13] Whether the position can even be affirmed without self-contradiction, I leave for another occasion. See also *Concepts and Society*, note 9 above, chapter 5.
[14] Erving Goffman, *Asylums* (Doubleday Anchor, New York, 1961).
[15] *Pace* Laing and Szasz.
[16] I argued this at length in *The Revolution in Anthropology* (Humanities Press, New York, 1964; Regnery, Chicago, 1968); and *The Story of Social Anthropology*, (McGraw-Hill, New York, 1972).
[17] Leon Festinger, H. W. Riecken and Stanley Schacter, *When Prophecy Fails* (University of Minnesota Press, Minneapolis, 1956). See also *The Revolution in Anthropology*, note 16 above, and Bryan Wilson, *Magic and the Millennium* (London, 1973).
[18] Barbara Goodwin, *Social Science and Utopia* (The Harvester Press, Brighton, 1978).
[19] See K. R. Popper, *Conjectures and Refutations* (Routledge and Kegan Paul, London, 1963), chapter 1.
[20] See Ernest Gellner, *Cause and Meaning in the Social Sciences* (Routledge and Kegan Paul, London, 1973), chapter 4.
[21] Peter Winch, *The Idea of a Social Science* (Routledge and Kegan Paul, London, 1958); A. R. Louch, *Explanation and Human Action* (University of California Press, Berkeley and Los Angeles, 1966); Keith Dixon, *Sociological Theory* (Routledge and Kegan Paul, London and Boston, 1973).
[22] Classically set out by K. R. Popper in *The Logic of Scientific Discovery* (Basic Books, New York, 1959).
[23] Harold Garfinkel, *Studies in Ethnomethodology* (Prentice-Hall, Englewood Cliffs, 1967).
[24] This seems to me to happen to the radical programme in the sociology of knowledge. See *Philosophy of the Social Sciences* 11 (1981), 173–243.
[25] Consider the disappointment over Headstart that led Moynihan to propose "benign neglect"; over crime and penal problems that led to 'labelling theory'; over Project Camelot that led to suspicion of all academic connections to government and so on.
[26] Talcott Parsons, *The Structure of Social Action* (Free Press, Glencoe (Ill.), 1937). See also B. Berelson and G. Steiner, *Human Behaviour: An Inventory of Findings* (Harcourt Brace, New York, 1964).
[27] *The Structure of Scientific Revolutions* (University of Chicago Press, Chicago, 1962).
[28] His credentials are scrutinised in Robert Merton, 'The Sociology of Science, An Episodic Memoir', in Robert Merton and Jerry Gaston (eds.), *The Sociology of Science in Europe* (Southern Illinois University Press, Carbondale and Edwardsville, 1977). Some of my comments are to be found in 'Laudan's Problematic Progress and the Social Sciences', *Philosophy of the Social Sciences* 9 (1979), 484–97 and my review of Kuhn's essays, *The Essential Tension*, in *Queen's Quarterly* 87 (1980), 65–8.
[29] Especially Robert Friedrichs, *A Sociology of Sociology* (Free Press, New York, 1970).
[30] J. Agassi, 'Scientific Schools and their Success' in his *Science and Society*, (D. Reidel, Dordrecht, 1981).

SPIRO J. LATSIS

THE ROLE AND STATUS OF THE RATIONALITY PRINCIPLE IN THE SOCIAL SCIENCES[1]

CONTENTS

1. INTRODUCTION: THE RATIONALISTIC APPROACH

How can we explain and predict human behaviour and, in particular, behavioural regularities in a social setting? Adequate answers to questions of this type have been of interest to influential traditions in the social sciences and in particular to sociology, economics, anthropology, and psychology.[2] The main concern of the present paper is a particular type of answer – let us call it a 'rationalistic answer' – to the above question. Rationalistic answers have been especially influential in economics, where an elaborate theoretical structure was built on a particular conception of human behaviour. The coherence, elegance and apparent problem-solving ability of the theoretical framework of neoclassical economics is probably responsible for the longevity and relative immunity of the rationalistic approach from incisive criticisms which have been levelled against it in the last few decades.[3]

The rationalistic approach goes something like this: 'We explain and predict behaviour in terms of a deductive schema which contains the agent's preferences, goals and objectives, an analysis of his situation, and the general assumption that agents behave adequately or appropriately to the situation.' This latter assumption is associated with such social scientists and philosophers as Pareto, Schutz, Popper and Talcott Parsons, and is sometimes known as the '*rationality principle*.'[4] It should be obvious by now that I am *not* referring to the 'Cartesian rationalism' of Hobbes and J. S. Mill but to a more sophisticated version which in part derives from Adam Smith.[5]

The rationalistic approach is supposed to go hand in hand with an

R. S. Cohen and M. W. Wartofsky (eds.), Epistemology, Methodology and the Social Sciences, 123–151.

individualistic approach to the explanation of social phenomena whereby social phenomena are *brought about and should be explained in terms of* individuals with aims, objectives and knowledge acting in a social situation which may contain institutions, traditions, systems of behaviour rules, etc. In this approach social phenomena are often the undesigned results of collectivities of individual actions in a social framework.[6]

In this paper I propose to criticise both the *individualistic* and the *rationalistic* components of this general approach. More specifically I shall claim that individualistic prescriptions should not be followed to the exclusion of other approaches to social research. I shall further argue that for an adequate account of an extensive range of behaviour, social theory should shed its rationalistic blinkers. My main criticism of the rationalistic approach is that it denies theoretical limelight to theories of human decision processes both at the individual and at the group level. In a world seen through rationalistic spectacles human behaviour consists of discrete steps of rational appraisals of situations, by a stylized theoretical agent. Institutions, traditions, habits, decision rules, heuristics, etc., are objectified aspects of an agent's situation. They are coolly and rationally calculated in the same way that sacks of rice are weighed and counted before being sold.[7]

My starting point however will be to ask whether our initial question concerning human behaviour and its social effects is the 'right' one, whether or not it is fruitful, or whether it is perhaps misconceived and misguided. I shall discuss this important problem in the next section and will conclude that, as a matter of fact, both holistic and individualistic questions may be asked and answered in a coordinated and complementary fashion. The starkly individualistic reconstructions offered by some economists and social philosophers are unrepresentative of what actually goes on. And the opposition of such reconstructions to extreme versions of holism (which nobody appears to have actually held, let alone practiced) seems philosophically unilluminating and theoretically unfruitful.

The status and role of the rationality principle will then be considered in some detail. In particular I shall critically appraise Popper's unsatisfactory treatment of the principle. I shall then develop what I consider to be the ontological underpinnings of the rationalistic approach using as my guide Popper's paper 'Of Clouds and Clocks' [1966]. Equipped with this background it will be easy to perceive human behaviour through rationalistic spectacles as *neither caused* by other events *nor random* and haphazard but as *plastically controlled* by situational analyses. Plasticity of control is, as we shall see, a matter of degree. Different degrees of plasticity, I shall argue, might require different explanatory approaches. The notoriously elusive and puzzling rationality principle then becomes palatable, for it is there to fulfil two roles: (1) to save the ideal of deductive explanation by providing logical validity to rationalistic explanations of behaviour, (2) to function as a 'plastic' as opposed to 'cast-iron' interface between an agent's situational scheme and the 'real' environments.

Next, I turn to the important gap between an agent's problem situation and his option-choice. This is the crucial part of the explanation schema, *not* the conceptual/'synthetic' gap bridged by the 'rationality principle'. The general shortcomings of the rationalistic approach become strikingly obvious when we consider the sterility it promotes in theorising by collapsing the situational picture, decision process and option-choice into a single, unproblematic step.

My conclusion points to the need for a change in emphasis and in outlook. On the level of individual behaviour we should place emphasis on the organisation and properties of human decision processes instead of considering them as inert situational features. The same behavioural questions that the rationalistic approach asks of individuals may be asked of complex structures of which individual agents are relatively unimportant parts. The general principles that govern the interactions and properties of such structures may be seen as central both to macroscopic social phenomena and to the regularities of behaviour of individual actors. Our tendency in the past has been to regard or to interpret social structures as 'ordered', 'stable' and 'coordinated' and to seek to discover those characteristics of the elements of such structures which contribute to stability and self-regulation. This for instance has been the case in economic theory with the notable exception of Keynes. In recent years this pattern seems to be changing – even in economic theory. Theorists, especially those interested in the foundations of economic theory, have started asking questions about instability and miscoordination and inquiring into the characteristics of social structures that contribute to this type of behaviour.[8] But while our theoretical interests seem to be shifting, the foundations of 'respectable' economic theory have remained unaffected.

2. HAVE WE GOT HOLD OF THE RIGHT PROBLEM?[9]

In the next few sections I shall be dealing with what I have called the rationalistic approach to the explanation of individual social behaviour. Yet certain justified doubts may be expressed concerning the value of this undertaking.

In particular the criticism may be levelled against me that I am trying to solve a relatively unimportant problem (for the purposes of explaining social phenomena) which, in any case, stands a very poor chance of being solved. Rather, it may be argued, all we can do in the social sciences is to predict the recurrence of complex patterns with their accompanying qualitative properties, i.e., we can at best predict that a social pattern of type K will in environment E exhibit properties $P_1, P_2, P_3, \ldots, P_i, \ldots, P_n$. For this we often need practically no explanatory theory of the behaviour of the elements because, by and large, behavioural regularities or irregularities do not affect the pattern. For want of a better term I shall label this latter approach '*systemic*'. I prefer to avoid the terms 'functionalist' or 'holistic'. The former is often seen to imply that the parts of an organised structure are needed to preserve some kind of 'order' while the latter is often associated with well-known, so-called 'holistic' theories of social and economic development.

There has been considerable and unnecessary controversy in the philosophy of the social sciences about which is the correct approach: the individualistic/rationalistic approach or the systemic one? It is in my view a mistake to emphasise the one approach to the detriment of the other because they can and do play complementary roles. As I shall try to suggest, in what follows, neoclassical economic theory is a good illustration of a social theory which embodies elements both of a *systemic* and an *individualistic* approach, although economists and methodologists have traditionally overemphasised the latter.

Let me say a little more about each approach before explaining why I feel they are reconcilable. I start with the individualistic/rationalistic approach and ask how it would deal with the question: 'Why has social event (or process) E occurred?' The answer usually goes something like this: 'E is the direct, intended or unintended outcome of the behaviour of one or many (usually many) individuals acting in a social situation. Given similar circumstances (in the inner and outer environment) and some assumption that behaviour in particular circumstances falls under some general rule[10], we should expect the same type of behaviour to occur and the same type of social effects to ensue.' The explanandum here is not an item or sequence of behaviours nor is it a behavioural regularity. But it is the direct outcome of a behavioural uniformity which may in turn be rationalistically explained. The uniformity in question is the main explanatory/'causal' factor in the rationalistic schema.

So the individualistic/rationalistic approach is not restricted to investigating behaviour or behavioural regularities but also takes an interest in the *mass effects*, intended or unintended, of a great many individual behaviours that exhibit some uniformity or constancy. Yet, the relational structure and properties of this mass of individual behaviours are not central explanatory factors and are hardly mentioned at all. That is, the explanatory schema does not contain organised structures which display some characteristic macroscopic behaviour and properties. In the individualistic/rationalistic approach, institutions, traditions, systems of behaviour rules and the like are treated as potentially analysable in individualistic terms and as appearing in individualistic explanations in the form of *obstacles to behaviour*. So it seems that structures or systems or wholes are acceptable in explanations of social phenomena provided they have the character of initial conditions or constraints. Systems do not 'behave' and do not possess general relational properties which explain social phenomena. They are *features* of a behaving individual's environment in the same way as a boulder blocking a mountain path is an obstacle to a climber. They are assessed rationally and behaviour appropriate to the individual's goals is initiated.

The *systemic* approach arises not so much from a different way of posing the problem but rather from a different way of approaching the solution. A social phenomenon rather than being viewed as the mass effect of a multitude of independent activities is now seen more like a *pattern of relations* (or their effects) where the elements composing the pattern are the unimportant and often irrelevant components of the pattern. In some cases the properties, values and behavioural characteristics of the elements are entirely irrelevant. We may, for instance, wish to claim that a 'flat hierarchy' 'learns' more quickly from the environment yet has lower chances of 'survival'. The truth or falsity of this claim is very largely independent of the kind of elements that constitute the hierarchically organised pattern. In other cases we make a general and vague assumption about the elements of the pattern. For instance if the elements of the pattern are economic agents we might assume that 'they will normally prefer a larger return . . . to a smaller one.'[11] It is important to notice that in the systemic approach the structures, systems and patterns we create have an existence almost separate from that of their elements. They exhibit behaviour, they interact with an environment and they exhibit well-adapted or maladapted characteristics. It is only a step away to claim that as these structures are moulded by the environment they change, changing with them or 'determining' the behaviour of their elements. But I do not wish to pursue here the claim that selection operates on structures or organised systems (or at least on groups of operations) as well as (or instead of) operating at the level of the individual organism.[12] For our purposes it is sufficient to convey that systems or patterns are viewed neither as sums of individual parts readily analysable into their components nor as inert structured obstacles in the path of a rational actor. I am satisfied with this negative way of distinguishing the two approaches as I do not believe there exists or should exist a sharp demarcation criterion. All the more so, as I want to suggest that both approaches could fruitfully coexist.

Neoclassical economic theory offers a good illustration of how a systemic interpretation may be used to 'patch up' what are widely considered to be paradigm cases of individualistic explanations of economic events. For instance when the empirical adequacy of some alleged regularity of economic behaviour is questioned we are told that nobody is really interested in predictions of particular economic actions or reactions.[13] What we are interested in is the structure and behaviour of the resulting pattern. In some cases, provided this pattern exhibits relations of a certain kind, then it will have certain properties, however diverse the behaviour of its constituent elements. In other cases some assumption concerning the elements of the pattern must be made if 'orderliness' is to be preserved. For instance, the assumption that 'economic agents prefer more to less' or that they are 'rational' is allegedly required to ensure that the economic system is 'orderly' and exhibits self-regulating tendencies.

Ever since Adam Smith, some of the central questions of economics invite – in part at least – a systemic treatment. Axel Leijonhufvud questions in a colourful and suggestive way some of the general features of economic organisations:

> The consumer wants milk in the morning. It is there on his doorstep, having arrived from a hundred miles away. The farmer milks his cow and has a consumer for it that he has never met a hundred miles away . . . How come shoes do not pile up unsold in New Mexico while people queue barefooted for shoes in Maine? In some parts of the world, such an event would not be all that unlikely . . . The economist who finds it wondrous will ask: "How is it possible – how is it even conceivable that decentralized economic activities can ever be reasonably coordinated when nobody, really, is trying to ensure the outcome . . ?" That I believe is the "right question".[14]

A starkly individualistic answer to this type of question – though logically possible – seems to miss the point entirely. The sort of problem that develops is of the following type: How do activities organised or interrelated in a certain way, coordinate? Would all patterns of elements or activities displaying the same relational structure also coordinate? Under what conditions should we expect this pattern or system to be stable and under what conditions unstable? As Leijonhufvud puts it, we want to see how a collectivity of interacting activities 'meshes' rather than to consider the mass effects of a multitude of independent activities.[15]

F. A. Hayek, in a series of important papers,[16] recognises (though not explicitly) the interplay between systemic and individualistic regulative principles in the explanation of social phenomena. Hayek's guiding problem is to explain 'orderliness' in social life – and in particular orderliness which has *not* been imposed or designed, but which is 'spontaneous'. This problem, Hayek claims, immediately places us in the domain of complex phenomena[17] where we must abandon the expectation of explaining and predicting singular events. At most we might be able to explain and predict abstract patterns of relations. The activities of human agents, being elements of the abstract patterns, are often required to exhibit some regularity or constancy, if order is to be preserved. That is, agents must be assumed to respond to their immediate environment – according to certain rules which have undergone some evolutionary selection process. The agents need not and often do not 'know' the abstract rules that they are following and have no knowledge of the character or effects of the interrelations holding among the multitude of individual activities. A Hayekian agent is often in no better position to appreciate the pattern of which he is a part than an iron filing is able to appreciate the role it plays in forming a pattern of 'force lines' constituted by a multitude of iron filings under the influence of a magnetic field.[18] According to Hayek " . . . these rules of conduct have . . . not developed as the recognised conditions for the achievement of a known purpose, but have evolved because the groups who practiced them were more successful and displaced others."[19] "We never act, and could never act, in full consideration of all the facts of a particular situation, but always by singling out as relevant only some aspects of it; not by conscious choice or deliberate selection, *but by a mechanism over which we do not exercise deliberate control*."[20] The systemic character of the 'spontaneous orders' Hayek describes becomes clear in the following passages. "All that is necessary to preserve such an abstract order is that a certain structure of relationships be maintained, or that elements of a certain kind (but variable in number) continue to be related in a certain manner."[21] " . . . the only possibility of transcending the capacity of individual minds is to rely on those superpersonal 'self-organizing' forces which create spontaneous orders."[22]

The picture Hayek presents retains only faint echoes of the individualistic/rationalistic approach. In Hayek's view rational appraisals of situations often give way to unconscious decision rules or behaviour rules. Moreover the discovery and reconstruction of these rules must be supplemented or even determined by the discovery of the character of the interrelations between the elements of the abstract order and the properties thereof.

Hayek's position – if I understand it well – reflects the kind of methodological approach I am adovcating here, i.e., an approach where we study complex phenomena by investigating the characteristic uniformities of the elements, the relational properties of the 'structures' or 'wholes' and the nature and consequences of the interaction between the two. I would, however, retain strong reservations concerning two of Hayek's theses about complex phenomena. One is his preoccupation with 'order', 'stability' or 'coordination' and the consequent attention focused on those relations, characteristics and properties of structures which are 'functional'. This approach detracts attention from dysfunctional characteristics or unstable relationships and seems unduly restrictive. My other doubt concerns Hayek's preoccupation with showing that in the social sciences we can hope for no more than the approximate explanation and prediction of abstract patterns. He – mistakenly – contrasts this to the possibility of 'explanation in detail' that has been 'realised' in the natural sciences. But I shall not argue against these theses at present as they do not substantially affect the argument of this paper.

What I have tried to illustrate is the systemic approach that a good deal of economic theory has (perhaps reluctantly) adopted. In a nutshell it is this: Instead of starting out by asking the question: 'How do human agents generally behave in economic situations?' it has asked the question: 'What are the minimal characteristics I need to attribute to the elements of a complex economic system so that the system as a whole displays a coordinated/self-regulated/orderly behaviour?' This is the sense in which I see 'systemism' and 'individualism ' as coexisting.

3. SUBJECTIVE OR OBJECTIVE SITUATIONS?

I now turn to the examination and appraisal of rationalistic explanations of social behaviour. Such explanations usually possess two characteristic features: the 'situation' in which behaviour takes place and some assumption of rationality or 'rationality principle' as Popper has called it.[23] One way of approaching the rationality principle is as the assumption that (for purposes of theoretical explanation) human social behaviour should be viewed as 'appropriate' or 'adequate' to the relevant circumstances or situation. But then the question arises: 'Whose circumstances or situation is rational behaviour appropriate to?' Is it the actor's conception of his situation or is it some objective notion of what were the relevant circumstances given the best available knowledge at the time the action took place (or was deliberated)? If it is the actor's view of his situation at some particular time, what is its relationship with the best available knowledge at that time? To put it another way, 'is behaviour to be termed 'rational' if it is appropriate to a subjective situation which is known to be false on the basis of the best information available to the actor?'[24]

Vilfredo Pareto was sharply aware of this problem and came down on the objective side, i.e., according to Pareto, rational action is that appropriate to an objective situation or to a subjective one that mirrors it. Pareto writes in his *Sociology* [1917]:

> There are actions which constitute means appropriate to some aim and which logically [or 'rationally'] combine towards this aim. There are other actions that lack this character. These actions are very different in accordance with whether they are viewed under their objective aspect or their subjective aspect. Under this first aspect practically all human actions [are rational]. For the Greek sailors the sacrifices to Poseidon and the activity of rowing were equally rational means of navigation . . . We shall call those actions rational which logically combine to attainment of the aim not simply from the point of view of behaving subjects but also from the point of view of those who possess wider knowledge.[25]

Talcott Parsons, in *The Structure of Social Action*, shares this view: "Action is rational in so far as it pursues ends possible within the conditions of the situation, and by the means which, among those available to the actor, are intrinsically best adapted to the end for reasons understandable and verifiable by positive empirical science."[26]

Popper – who to my knowledge first coined the phrase 'rationality principle' – weakens the notion of rationality in explanations of behaviour to something that closely resembles Pareto's "subjectively rational actions". He writes:

> We try to explain a madman's actions, as far as possible, by his aims (which may be monomaniac) and by the "information" on which he acts, that is to say, by his convictions (which may be obsessions, that is, false theories so tenaciously held that they become practically incorrigible). . . The rationality principle is a 'minimum principle' since it assumes no more than the adequacy of our actions to our problem situations as we see them.[27]

Popper is not entirely consistent on this. In other passages of the same paper he hints at a stronger version of the assumption of rationality which resembles the definitions by Pareto and Parsons. But then we would not be dealing with a 'minimum principle' but rather with a stronger rationality theory.

Before I discuss the more restrictive version where rationality approaches optimisation I will examine the rationality principle as a 'minimum principle'. It is easy to see that the rationality principle so construed lets through as rational most social, economic, political, problem-solving and even neurotic behaviour. This is so, because the agent's subjective situational analysis – to which behaviour is to be appropriate – need have no connection, or a very tenuous one, to the situation as an 'objective' outside observer (possessing some readily available knowledge) would have construed it at the time.[28] So the rationality principle is wider, embraces more, and is weaker compared to what we usually consider as criteria of reasonableness for human behaviour.

4. THE RATIONALITY PRINCIPLE: FROM OPTION-CHOICE TO ACTION

Having briefly considered some problems surrounding the notion of 'situation' we may now turn to the question of the status of the rationality principle and that of its 'role' in rationalistic explanations of behaviour. The discussion in this section will be based on Popper's writings and especially his [1967] paper referred to above. The main argument to be developed in this section may be put forward in three steps: (a) Popper's account of the role and status of the rationality principle is obscure and unsatisfactory. (b) A reconstruction of Popper's account leads us to view the rationality principle as a premise connecting or bridging statements describing option-choices and descriptions of behaviours. (c) But this – I believe plausible – reconstruction of Popper's views does not square with the role and properties he wishes to attribute to the principle.

I start with the role of the rationality principle in a Popperian explanation schema of behaviour which, as we have seen, consists of the *situational analysis* and the *rationality principle*. The situational analysis includes two sorts of things: Goals or aims, and knowledge or information. The real agent has drives, desires, wants, etc. In the Popperian schema these psychological features are objectified; they become, as Popper notes, "abstract situational features."[29] The hunger drive becomes a goal or an abstract aim to consume some edible substance. It may perhaps be represented by an indifference map – a device familiar to economists. The second component of the situational analysis is the agent's beliefs. This turns into abstract situational information, i.e., the theories and facts in the possession of the individual concerning his situation in its relevant physical and social aspects and including the means-technology the agent judges necessary for the attainment of his goal. The above components correspond to the *initial conditions* – Popper calls them "typical conditions" – of a deductive explanation of some physical event. On this view, the difference between physical and social events seems to derive from the degree of specificity attainable: In the case of the situational analysis we want to *simulate* behaviour via a 'rough and ready' model rather than to predict a spatiotemporally singular event.[30]

The next question Popper raises is this: How do the various elements in the explanation schema 'act upon each other'? Put it another way, which component of the model stands in for the 'law'? In social models, Popper asserts, the rationality principle stands in for the 'law' which 'animates' the otherwise inert collection of situational features. But, Popper warns us, its status is *not* that of a 'law' or of an empirical hypothesis. From here on Popper's treatment seems either confused or deliberately elusive. He says that the principle "corresponds", in social explanations, to Newton's laws of motion in physical explanations and yet the principle "does not play the role of an empirical theory". It is "almost empty" "not a priori valid", "clearly false", "a good approximation to the truth" and "the consequence of a methodological postulate". "Nevertheless", Popper writes, "my views on the rationality principle have been closely questioned: I have been asked whether there is not some confusion in what I say about the status of the principle . . ."[31] But Popper does not go on to explain how this diverse and, on the face of it, incoherent collection of properties may be plausibly attributed to the rationality principle.

Watkins in his *Imperfect Rationality* [1970*a*] has attempted to distil some coherent view of the properties of the rationality principle. Two closely related properties stand out: its *syntheticity* and its *indispensability*. Do these properties harmonise with Popper's characterisations? How can something be 'not empirical,', 'not *a priori*', not a principle of conceptual connection, and at the same time 'synthetic'? To try to make some sense of conflicting claims about the status of the principle it helps to go into its role in explanations of social behaviour in slightly greater detail.

Consider a theoretical agent in a problem situation which necessitates some overt response. The situational analysis, of the agent in question, which consists of aims and information, somehow 'works out'[32] the appropriate option-choice in the circumstances. (The option-choice seems to be considered 'implicit' in the situational analysis).[33] So the role of the rationality principle is *not* to 'tease out' of the situational analysis an option-choice or a 'practical conclusion'. Rather its role is to effect a connection or to bridge a gap between the decision to do something at t and the actual performance of the behaviour at t or to be precise, at $t + \epsilon$ where ϵ can be as small as we want it to be. The claim that this principle is neither one of conceptual connection nor analytic is based on an argument of the following type: The decision now to do X does not imply that I will actually now implement the decision. It is quite possible for me to decide now to do something and not do it even in the absence of any new considerations or psychological constraints that may interpose themselves between option-choice and action. And since the principle says something that could conceivably be violated – the argument goes – it must be synthetic. But then what sense can we make of the declarations that the principle has no empirical content, or at least practically none, and is neither physiological nor psychologistic?

To recapitulate, the rationalistic approach proposes the method of situational analysis in order to explain action in terms of its adequacy relative to abstract situational features and without resorting to psychological considerations. The role of the rationality principle seems to be that of a non-empirical, bridge-principle which nevertheless must say something about the connection between mental states and behaviour which could be violated though not refuted.[34]

Watkins makes a splendid effort to illustrate precisely this point. He takes his example from a decision situation in a novel by Iris Murdoch: "Dora, a young woman who has decided to rejoin her husband, gets to Paddington Station early and finds a seat on the train. It is a very hot afternoon and the train becomes very crowded, with many people standing. An elderly lady halts in the doorway of Dora's compartment: it seems that she is a friend of the person sitting next to Dora and that they had hoped to travel together. It occurs to Dora that she ought, perhaps, to give up her seat. She silently argues against this:

> She had taken the trouble to arrive early, and surely ought to be rewarded for this . . . There was an elementary justice in the first comers having the seats . . . The corridor was full of old ladies anyway, and no one else seemed bothered by this, least of all the old ladies themselves! Dora hated pointless sacrifices. She was tired after her recent emotions and deserved a rest. Besides, it would never do to arrive at her destination exhausted . . . She decided not to give up her seat. She got up and said to the standing lady 'Do sit down . . . '[35]

If we accept this illustration of the syntheticity of the gap, it remains difficult to square it with the accompanying claim that the principle that bridges it is devoid of psychological or social psychological content. Far from putting our mind at rest, the example serves to reveal the sort of hypotheses that the rationalistic approach excludes and which may 'determine' behaviour. Such are hypotheses concerning conscious or unconscious behaviour rules followed in situations of conflict or stress. Incidentally, Watkins considers Dora's behaviour as a "humorous portrayal of human inconstancy."[36] In general, the rationalist's reaction to situations of this type is to throw his hands up with charming helplessness and say: 'I can't explain it; it is one more example of human freedom, inconstancy and unpredictability.' But it is not at all clear that this need be the case.

To a large extent, whether or not we see a gap between option-choice and action depends on how we have designed and interpreted the explanation schema. That is, if the premises consist of an objectified situational analysis – which somehow 'throws up' some innate option-choice – then, by construction, there is a gap between option selection and action (though the gap could be conceptual and not synthetic). Notice, however, that if we had included in the premises a motivational assumption as understood in modern physiological psychology, then we would have been able to deduce behaviour without the interposition of either synthetic or conceptual connection principles. To see this more clearly, consider a crude example: Suppose that the motivational assumption attributed to the agent is some stimulus-response chain '*S-R*'. If to this we add, as an initial condition, the presence of stimulus circumstances '*S*', then behaviour follows from these two types of premises alone. But with Popper's objectified, situational concepts in the premises it is apparently impossible – by design – to get motor responses from situational premises, even if we are allowed to assume that a situational analysis always 'throws up' some option-choice. Still evaluations, selections, failures to act and mental acts of all kinds could presumably be generated by the situational analysis *without the principle*.

A subsidiary thesis often accompanies the argument for syntheticity. The crux of this thesis is that by virtue of not allowing (by methodological decision) the rationality principle to be falsified we actually increase the falsifiability and reduce the arbitrariness of our models.[37] For consider what would happen to the situational schema if we denied the rationality principle. We would break the common link between all possible situational reconstructions and the explanandum behaviour. Put it another way, the relative evaluation of alternative situational analyses that could have led to the behaviour becomes unnecessary. But this argument, attractive as it may appear at first sight, is only convincing to one who has 'bought' the 'rationality principle' package. It does not furnish additional arguments in support of the syntheticity and indispensability of the principle. For those who have not 'bought' the rationalistic recipes it is no great loss to be told that in certain cases – i.e., where we direct the modus tollens to the rationality principle – we might be deprived of the opportunity of comparing rival situational reconstructions. In Dora's case, for instance, we might explain behaviour by invoking an unconscious behaviour rule which operates in conflict and stress environments. And those who, like Donagan[38], consider the gap between the practical conclusion of situational analyses and action as being analytic are not saddled with arbitrary or *ad hoc* explanations. Their explanatory schemata will be as good as the empirical adequacy of the premises.

Several, mostly negative, properties of the principle reluctantly emerge:

(1) It refers to what I call the gap between option-choice and action, not, as perhaps our intuition would tell us, to the process of deliberation or decision.

(2) It does not have the character either of an empirical theory or of a principle of conceptual connection.

(3) It is an indispensable prerequisite for any explanation of social behaviour *so construed* and is therefore of no value in relative appraisal.

I do not believe that further and deeper investigation of the writings on the rationality principle will reveal much more than these negative and somewhat vague characteristics. I would moreover claim that this lack of concrete and positive suggestions concerning the status, role and content of the principle is *not* the result of neglect or oversight on the part of its proponents. In the next section I shall argue that the only way to understand its obscure role and vague status is to see it as a relic of a particular solution to a philosophical problem. More specifically, I shall argue that the rationality principle has been selected to bridge the gap between descriptions of mental states and actions *precisely because it says nothing about the process*. Knowing something general, 'law-like' or specific about the process would have been incompatible with a certain ontology about the relation between mental states and behaviour.

5. SITUATIONAL ANALYSES AS PLASTIC CONTROLS

In his well known paper 'Of Clouds and Clocks' [1966] Popper offered – among other things – a solution to an aspect of the so-called 'mind-body problem' concerning the manner in which mental states affect behaviour.[39] My account of Popper's solution is bound to be something of a caricature. It is much too brief to do justice to Popper's long and wide-ranging paper and it focuses on those aspects which serve my purpose, namely, to show how one part of Popper's philosophy may be made more intelligible by showing its harmony with another (even though the resulting ensemble in my view harbours undesirable methodological consequences for the social sciences).

But why is it necessary to bring Popper's ontology into a discussion of his approach to the explanation of behaviour? Am I bringing in a 'plastic herring'? Quite the contrary. Popper's [1967] paper presents a certain – rather obscure, as I have explained – approach to the explanation of social behaviour. But in an earlier and important paper [1966] Popper puts forward an ontology concerning the relation between mental states and behaviour. What could be more natural than to try to see if this earlier paper throws any light on the later one?[40] This is my justification for this perilous undertaking.

Popper's starting point is what he calls "the nightmare of the determinist". According to Popper, the nightmare goes back to David Hume, who claimed that behaviour is either *predictable and determined* or *random and unpredictable*. This situation has appeared undesirable to many philosophers because, among other things, notions like 'freedom', 'choice', and 'innovative behaviour' lose their customary meaning. Popper wishes to escape the ugly consequences of what he considers to be Hume's dilemma by developing an account of behaviour which is neither random nor determined but somewhere in between.

To this effect he introduces some terminology. He compares living and non-living systems to *'clouds'* and *'clocks'*. A 'cloud-like' system is a disorderly and unpredictable one like a cloud of gas. A 'clock-like' system is a regular, orderly and predictable one like the planetary system. If Hume had been using Popperian terminology he would have said that human behaviour is either 'cloud-like' or 'clock-like'. Popper wants behaviour to reside in the area between the 'clouds' and the 'clocks'. A satisfactory – for Popper – account of the interaction of mental states and behaviour is neither analogous to the interaction between the molecules of a gas nor analogous to that holding among the various parts of a clockwork mechanism. To solve the problem Popper interposes 'soap-bubbles' between the 'clouds' and the 'clocks'. He also introduces the idea of degrees of control – the idea that systems exert 'a kind of action'[41] upon their elements. A 'clock-like' system exerts 'absolute control' over its elements. A 'cloud-like' system exerts 'no control' while a 'soap-bubble-like' system exerts *'plastic control'* over its elements.

Popper is unfortunately not clear about the nature and operation of plastic control. But we are fairly clear about the problem that the notion of plastic control is addressed to. As Watkins [1975] puts it, it is part of the answer to the question: "Do we have a model which could account for the exercise of a degree of influence or control by *A* over *B* where *A* is a man's mind and *B* his body?"[42] Schematically, we might place 'plastic control' on a line with 'no control' at one extreme and 'cast-iron control' at the other (*Schema I*), and with 'plastic control' somewhere in between.

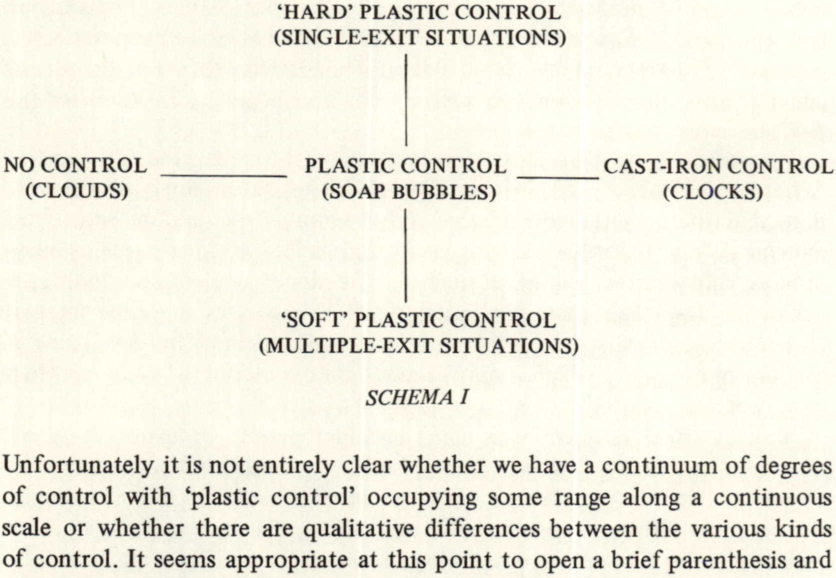

SCHEMA I

Unfortunately it is not entirely clear whether we have a continuum of degrees of control with 'plastic control' occupying some range along a continuous scale or whether there are qualitative differences between the various kinds of control. It seems appropriate at this point to open a brief parenthesis and attempt to bring together Popper's scattered remarks on plastic control.

We start by identifying the *object* of control and the *means* of control. The objects of control seem to be all purposive human behaviour and even some nonpurposive behaviour or behaviour guided by unconscious rules.

The means of control may be 'aims or purposes', 'proposals', 'problem-solving methods' and 'critical arguments'[43] – presumably including theories and their critical appraisals. The crux of Popper's compromise solution is that our mental states control some of our behaviour and that this control is "of a 'plastic' kind".[44] But what does 'plasticity' of control imply? Popper is not clear on this. The closest he comes to discussing the nature of 'plastic control' (apart from likening the set-up to a soap bubble) is when he suggests that we are not forced to submit ourselves to the control of our mental states but are 'free' either to alter them or, at least, not to act in accordance with them. The other aspect characteristic of plastic control is that it is control with feedback, or, as Watkins puts it, that 'there is a certain give and take between that part or aspect of the system which is more controlling than controlled and those parts which are more controlled than controlling'.[45]

Our previous discussion of the rationalistic account of behaviour may now be reformulated as follows: Decision behaviour is plastically controlled by mental states – among other things. These mental states are composed of aims, purposes, theories and critical arguments and may be represented in terms of what we have called the situational analysis. Any 'situational analysis' may thus be viewed as a set of 'plastic controls', i.e., as something that tends to regulate behaviour but does *not* force the behaving organism to comply to a strict behavioural repertoire.

It is now, I hope, easier to see how Popper's ontology is related to his discussion of the methodology of the social sciences. Any situational analysis may be seen as consisting – in part at least – of a set of conjectures that an organism holds about its environment. These conjectures interact with the environment and even change as a result of the 'give and take' relation with the organism's surroundings. An actor may therefore be viewed both as controlling and as being controlled by his situational analysis. And, of course, 'lower-level' situational analyses may be plastically controlled by higher ones and so on.[46] Consider any behavioural event of the form '*A* does *X*'. And let us ask the question: 'How have the antecedent mental states (or objective situational analyses) affected it?' According to the view presented in Popper's [1967] we should consider the abstract features of *A*'s situational picture. This presumably means that drives or motives become abstract aims or objectives and that beliefs become (relevant) situational information. Still, in order to deduce the explanandum event from the situational analysis we must add something to the premises. But if we fill the gap with some universal statement concerning the manner in which drive states (or a psychological theory of motivation) 'cause' or 'affect' behaviour, are we not back to the Humean 'clock-like' interpretation of the relation between mental states and behaviour? Paraphrasing Popper, we may call this the 'indeterminist's nightmare'. That is, how can we retain deductive predictability and yet *avoid* proposing an explanans which suggests 'cast-iron' control between mental states and behaviour? Or, to put it another way, how can we reconcile the philosophical theory of plastic control with the thesis that we may deductively explain and predict social events including human actions and their effects. The 'conciliator' is the 'rationality principle', and the 'indeterminist's nightmare' in my view explains Popper's obscure position with respect to the role and status of this principle. For instance, we can see why it has neither the status of a universal theory nor that of a translation principle. It does not have the status of a universal theory because Popper's ontology does not allow him to represent the connection between mental states and behaviour as a causal one. But it is not a translation (or conceptual connection) principle either, because it *does* say something (or, at least, we might have liked it to say something) about the way in which mental states and behaviour are connected. The rationality principle is the interface between the world of mental constructs which plastically control behaviour and the physical world that constrains and moulds it. The role of the rationality principle is to function as a '*plastic interface*' between *mental states* and *behaviour*. This is the reason why the rationality principle is 'false but close to the truth'.[47] It is false if taken literally because it *does not determine behaviour*. But it is close to the truth because it, along with the situational analysis, represents the *tendency to control behaviour* implicit in Popper's ingenious ontology. The rationality principle has no specifiable content just as the thin layer of molecules that constitutes the surface of a soap bubble has no specifiable shape. Slight variations in temperature may cause it to expand or contract (or explode) and contact with other surfaces may alter its shape. It is what it is because of the forces from within and those from outside. And yet it is indispensable in the sense that it makes the soap bubble what it is by connecting the inner with the outer environment.[48]

6. THEORIES OF THE DECISION PROCESS: FROM PROBLEM SITUATION TO OPTION-CHOICE

I now want to turn from the rationality principle to descriptive rationality theories or 'theories of the decision process' as I prefer to call them.[49] The rationalistic approach typically conflates these two crucially different types of bridging premises. The briefest and simplest way of expressing the difference is by noticing that while the rationality principle connects option-choices with actions, a 'theory of the decision process' or a 'decision rule' connects what we may call a 'problem situation picture' with the resulting option-choice. I am using the phrase 'theory of the decision process' in the widest possible sense. Thus, the hypothesis that agents follow certain possibly unconscious or innate decision rules or systems of rules is in my terminology a hypothesis concerning the decision process. So is a search and cost-benefit analysis of the alternative options of a complex project. So are hypotheses concerning the dynamics of goal or preference formation.

The advocates of the rationality principle have very little to say about decision processes. Popper seems to suggest that an option-choice lies implicit in the situational analysis or that the situational analysis somehow 'works out' the appropriate option-choice. Put it another way, in the rationalistic approach *there is no explicit decision process*. Yet, there is little reason to suppose that from goals, objectives and situational information alone, some option-choice – or even a set of option-choices – will emerge.

Pareto's assumption of rationality, on the other hand, does seem to be richer than the stark rationality principle. In particular it suggests that an optimising selection rule operates, i.e., on the basis of objectively good situational information, agents always select that option that could not be bettered in the circumstances. In his earlier writings and in particular in his *Poverty of Historicism* [1957] Popper seemed to adopt a somewhat similar assumption of rationality which he presumably relinquished as too strong for the purpose of his [1967] paper. If we accept, as seems reasonable, that some general statement indicating how problem situations and option-choices are connected, is either explicitly or implicitly involved in explanation schemas of behaviour, then the question of its status arises. Is it to be treated as falsifiable or as *a priori*? Is it to have the character of an empirical theory or that of a translation principle? What is the status of, say, the optimising decision rule? The characterisations 'empty', 'trivial' and the 'consequence of a methodological postulate', clearly no longer apply. On the contrary, an optimising theory of the decision process has empirically criticisable consequences and underlies elaborate social theories. It would therefore seem appropriate to treat such hypotheses in the same way as we would any other empirical hypothesis. But would we not then encounter difficulties analogous to those we would expect to meet if we treated the rationality principle as an empirical law? For we may be accused of creating a Humean 'clock-like' explanation schema which *determines* option-choices, as opposed to 'plastically controlling' them. Put it another way, problem-situations or situational analyses would then affect option-choice in a clock-like manner. And this is no good to the indeterminist because it represents mental acts as 'determined' while their physical accompaniments are 'free'. This analysis might explain Popper's reluctance to spell out relevant decision rules and his insistence that the appropriate option somehow 'falls out' of the situational analysis. And yet Popper's position may not be as *ad hoc* as it appears at first. Certain situational analyses seem to have a compelling or severely constraining character. Aristotle's example of somebody 'throwing the cargo overboard in a storm'[50] or the behaviour of a seller under perfectly competitive conditions, readily come to mind. Though I may be straining Popper's 'plastic' terminology beyond the breaking point, I am tempted to say that some problem situations or situational analyses exert 'hard' plastic control over option-choices (see *Schema I*, p. 138 above). Such situations (which I have elsewhere called 'single-exit' situations[51]) do not formally dispense with the need for a bridging decision rule. But they do make it more plausible to believe that the situational analysis somehow (with 'very little help') 'works out' an option-choice. Moreover, in a compelling situation of the type described above, a variety of alternative decision rules would generate identical option-choices. The same cannot be said about the other end of the spectrum where we have 'soft' plastic control. Imagine an organism placed on a vast plane surface whose properties do not immediately affect the organism's essential variables. The problem situation (if we can call it that) imposes very 'soft' plastic control indeed on the option-choices. (Eventually we would expect thirst and hunger to create a problem situation. Yet, the environment, as we have described it, presents the agent with an 'open' or multiple-exit situation). In this case the *decision process* rather than the situational analysis will generate the option-choice. For instance, an agent faced with an investment portfolio decision would probably be 'controlled' by a fairly 'soft' situational analysis and so would a seller making a pricing decision under oligopolistic conditions.[52]

7. CONCLUSION: SHORTCOMINGS OF THE RATIONALISTIC APPROACH

The rationality principle and the accompanying rationalistic approach leave the explanation of behaviour devoid of what Aristotle has called 'deliberation' or what we have referred to here as 'decision processes'. There seem to be good reasons to suppose that some types of social environment give rise to behaviour that requires only token or trivial deliberation. I have referred to such situations as exercising 'hard' plastic control on the behaving organism. The rationalistic approach becomes quite palatable for the analysis of behaviour under such conditions. Aristotle, in his *Nichomachean Ethics* recognises closely analogous cases " . . . where there is exact and absolute knowledge, there is no reason for deliberation; e.g., writing: for there is no doubt how the letters should be formed."[53] And Aristotle goes on to say: "Matters of deliberation, then, are matters in which there is an element of uncertainty."[54] As we move across the spectrum from highly constraining situations (which narrowly delimit behavioural alternatives), towards open situations where option-choice does not 'fall out' of the situational analysis, the model of a decision maker without a decision process becomes increasingly implausible. Put it another way, the 'rationality principle' approach is more likely to succeed in those areas where choice or behaviour is *independent* of the process by which it is reached, i.e., in situations where the agent is subject to 'hard' plastic control. The 'rationality principle' approach is inappropriate where choice or behaviour is *dependent* on the process by which it is reached, i.e., where there is 'soft' plastic control.

The contrast I have been trying to emphasise bears close affinity to Herbert Simon's distinction between 'substantive' and 'procedural' rationality. 'Substantive' rationality is that appropriate to situations where there is 'hard' plastic control – the rationality of 'economic man'. 'Procedural' rationality on the other hand involves consideration of decision processes and cognitive processes and seems more appropriate to the other end of the spectrum.[55]

Popper's writings on the rationality principle mistakenly represent a philosophical reconstruction of one research programme whose main domain has been economic theory as *the* method of the social sciences. But leaving that aside, I do not see how we can bring direct *a priori* or empirical arguments to bear against the rationalistic approach (or any other metaphysical research programme for that matter). To be sure we may point to the increasingly diminishing returns the approach is getting within micro-economic theory.[56] But even this is no guarantee that with sufficient ingenuity we could not revive it or apply it to another field like anthropology or cognitive psychology.[57] Yet if we suspect, as I do, that situations where there is 'soft' plastic control abound in social life, then we could point to the restrictiveness of the rationalistic approach; that is, we could point to those types of general propositions that would be repelled by the 'negative heuristic' of the approach.

I have lumped a considerable amount of alleged theoretical material under the label 'theories of the decision process' but have said very little about what we find when we peel off the label. I want to end by giving some brief – and hardly adequate – examples of what I mean by 'theories of the decision process'.[58]

To start with we should notice that there are alternatives to the optimising decision rule like the 'satisficing' and the 'cost-benefit' decision rules which, unlike the optimising rule, allow room for decision processes. According to the 'satisficing' decision rule, put forward by Herbert Simon [1955], instead of attempting to explain the behaviour of agents as best decisions in a determinate situation, we should attempt to explain them as more or less good (or possibly disastrously bad) solutions in fluid and partially known or even completely misunderstood situations. Moreover, the 'satisficing' approach does not restrict itself to fixed goals but embodies flexible aspiration levels which adjust upwards and downwards with experience. The 'decision process', like the Aristotelian notion of deliberation, involves all the steps from the vague glimpse of a problem situation to the resultant option-choice.[59] Hypotheses concerning 'search' and 'information gathering rules' are also involved in what I call the 'decision process'. Recently Tversky and Kahnemann have come up with interesting hypotheses concerning the heuristic rules unconsciously employed by problem-solvers in information gathering. One of these rules, *'adjustment and anchoring'*, is described as follows:

> In many situations, people make estimates by starting from an initial value which is adjusted to yield the final answer. The initial value or starting point may be suggested by the formulation of the problem . . . Whatever the source of the initial value, adjustments are typically insufficient. That is, different starting points yield different estimates, which are biased towards the initial values.[60]

A closely related general criticism is that the rationalistic approach seems to exclude, or at least to discourage from its research policy, the explicit analysis of human interaction as this is understood in social psychology. According to the proponents of the rationality principle interaction between *A* and *B* is the action of *A* holding *B* as an abstract situational feature and vice-versa. This tendency in turn discourages the production of potentially fruitful explanatory hypotheses of, say, 'conflict reduction' or of 'behaviour under conditions of threat.'[61] For instance, such hypotheses would be central to the explanation and prediction of bargaining behaviour where interaction during the decision process is of primary importance.

Let me end on a conciliatory note. The rationalistic approach to the explanation of behaviour has generated an important and pervasive research programme in the social sciences. As I have tried to show, it derives (among other things) from a certain ontology concerning the connection between mental states and behaviour. Like any research programme it excludes a great deal, idealises a lot and moulds – sometimes misleadingly – our view of social reality. I have suggested that for a certain range of problems its application may be fruitful while for another range its limitations appear overwhelming.

NOTES

[1] An abbreviated version of this paper was presented to the Boston Colloquium for the Philosophy of Science on February 24, 1976. See also Section 2(b) of my [1972] for a less detailed analysis. The present paper was submitted in July 1976 and has not been subsequently revised.

The author wishes to acknowledge helpful criticisms by Charles Christenson, Noretta Koertge and John Watkins.

[2] I am not of course suggesting that such questions are the *only* or *most* fruitful ones to ask in the social sciences. See especially Section 2 below.

[3] Cf. for instance Simon [1955] and Rothchild [1947] for two important critiques of the rationalistic approach and its consequences.

[4] Both Pareto and Popper remark that most human behaviour is 'non-logical' or 'non-rational' but add that for the purposes of explanation and prediction we should consider it 'as if' it were 'logical' or 'rational'. See Pareto [1899], Section 149 and [1917], Chapter III, Section 1 and Popper [1967].

[5] See Hayek [1973], especially Chapter 1.

[6] Cf. Popper [1945], [1957] and Watkins [1952].

[7] We must not of course overlook the fact that the denial of an explanatory role to decision rules, information gathering rules, heuristic rules, etc., may in part stem from the need to impose some single uniformity on the vast diversity of human social behaviour. My procedure will be to sympathise with this need for drastic simplification and to examine its effects. I therefore part company with those who have attacked the rationalistic approach by appealing to its implausibility or descriptive inaccuracy. A fuller discussion of the problem would include an appraisal of the 'theoretical output' of the rationalistic approach. However, as I have gone some way in this direction in my [1972], [1974], and [1976*b*], I shall not duplicate the material here.

[8] For this story see Leijonhufvud [1968] and [1976] and Loasby [1976].

[9] The present section is in a sense an apology for separating 'systemic' and 'individualistic' questions in the remainder of the paper. For an illuminating account of the questions economic theory has traditionally asked – as well as for advice on what they should be – see Leijonhufvud [1976].

[10] Cf. below, Sections 3 and 4.

[11] Hayek [1973], p. 45.

[12] Cf. Alexander [1975].
[13] Cf. Machlup [1946], [1952], [1967] and [1974] for the most consistent and well-argued formulations of this position.
[14] Leijonhufvud [1976], p. 89.
[15] There seems to be a close affinity between this type of question and that asked by Ashby of the organisation of the brain considered as an adaptive mechanism: "How then do the activities of the neurons become co-ordinated so that the behaviour of the whole becomes better, even though no absolute criterion exists to guide the individual neuron?" (Ashby [1952], p. 7).
[16] Hayek [1967], Chapters 2, 3 and 4. For further developments of these ideas see Hayek [1973].
[17] Hayek [1967], Chapter 2.
[18] Hayek [1973], p. 40. That this can be so becomes obvious when we consider the linguistic behaviour of children who are capable of using a language adequately without 'knowing' or being capable of articulating the relevant rules of grammar and syntax. On this see Hayek [1967], Chapter 3.
[19] Hayek [1973], p. 18.
[20] *Ibid*., p. 30. *My italics.*
[21] *Ibid*., p. 39.
[22] *Ibid*., p. 54.
[23] Popper [1967].
[24] The attribution of 'rationality' to behaviour, theories, designs or appraisals is often taken to signify approval. And beyond approval the characterisation of behaviour as 'rational' often carries recommendatory force. But besides this *normative* aspect rationality embodies a *descriptive* dimension. According to Hempel to qualify an action as 'rational' is to put forward an empirical hypothesis of the form: "The action was done for certain reasons and may be *explained* as having been motivated by them" (Hempel [1965], p. 463. His italics).

Social scientists ideologically as far apart as von Mises and Myrdal claim in their [1949] and [1954] respectively that the concept of rationality embodies inseparable descriptive and normative elements. I shall dissent from this view in the present paper because it is centrally important for a critical empiricist methodology to be able to single out the descriptive element of a theoretical system, i.e., that part of the system which is directly accountable for empirical discrepancies.

In general there are two ways in which rationality may be a normative principle. The first is damaging to an empiricist methodology; the second innocuous. In the first sense the principle recommends that all actors *should* (*although they may not*) act in such a way as to satisfy some criterion. On this conception of rationality some or all actors may behave as a *matter of course* in a way that does not satisfy the rationality criterion. This empirical evidence would, of course, leave the normative rationality criterion unaffected. The extent of empirical discrepancy would only be indicative of the number of deviants who must be 'kicked into line'. It follows that a conception of rationality which is normative in this sense may be totally inappropriate as an explanatory principle. The second conception of rationality embodies distinct normative and descriptive elements. The normative element is contained in the *criterion of selection* the actor himself imposes on the alternative courses of action open to him, given his goals and his situation. Any such criterion has normative force because it *recommends* to the actor the *appropriate* means available in his inner and outer environment which enable him to attain his goal. To put it another way, once the actor has accepted the criterion, then it has *normative force for him*. But this does *not* mean that any particular criterion or selection rule is recommended, for there may be an indefinite variety of such criteria. And, of course, which selection rules the economist or sociologist decides to apply must ultimately be an empirical matter too. Indeed it would seem quite preposterous to require that an outsider trying to understand the action in question should share the actor's rationality criterion and be compelled by its emotive force. From the point of view of the outsider who is explaining the action, the rationality principle is *descriptive*.

[25] Pareto [1917], Section 150.

[26] Parsons [1937], p. 58.

[27] Popper [1967], pp. 147–148.

[28] It should be noted that in order to explain unintended consequences of actions we may need to involve both subjective and objective situational features. Consider an example: I enter my home at night and flick the light switch. The light alerts a burglar who is trying to get through the window. The burglar is startled and falls off the window-ledge onto the head of a passing policeman. These unintended effects of my action, though in a sense directly deriving from it, *cannot* be accounted for purely on the basis of my aims, my subjective situational picture and the rationality principle. It becomes necessary to introduce into the explanation schema a more comprehensive or 'objective' account of the relevant environment.

[29] Popper [1967], p. 144.

[30] *Ibid*., p. 145.

[31] *Ibid*., p. 146.

[32] *Ibid*., p. 146.

[33] I shall have more to say about how a situational analysis 'works out' an option-choice when I discuss decision processes in Section 6 below.

[34] On this see Latsis [1972] and [1974].

[35] Watkins [1970*b*], p. 220.

[36] *Ibid*., p. 230.

[37] Popper [1967], pp. 146–147.

[38] Donagan [1970], pp. 218–227.

[39] Popper extended his [1966] ideas in his [1968*a*], [1968*b*] and [1974]. In his [1974] he puts the problem as follows: " . . . how can the relation between our bodies (or physiological states) and our minds (or mental states) be rationally understood?" (p. 149).

[40] Another proponent of the rationalistic approch – Ludwig von Mises – offers an explicit but rather crude dualist ontology. Unlike Popper, however, he closely and explicitly relates his ontology to his methodological prescriptions for economics. Cf. von Mises [1949].

[41] Popper [1972], p. 209.

[42] Watkins [1975], p. 214. Also see Watkins [1974], p. 397 where he contrasts the central control system of an organism with its motor system.

[43] Popper [1966], p. 239–241.

[44] *Ibid*., p. 252.

[45] Watkins [1974], p. 389.

[46] In fact, as Popper [1966] and Watkins [1974] suggest, organisms may be viewed as hierarchically organised systems of plastic controls.

[47] Popper [1967], pp. 145, 148.

[48] Some of the foregoing discussion might have been avoided if we could have claimed that the notion of 'plastic control' is totally unrelated to the degree of predictability. However, this does not seem to have been either Popper's or Watkins' position. (See, however, Feigl and Meehl [1974].) Popper writes characteristically: " . . . the fact that world 3 has a (plastic) control over much of our action shows, I think, that it has also some control over our brains; if this is so, then the essential unpredictability of world 3 enters world 1 along this path as well" ([1974], p. 1058). And Watkins adds: "If it is the controlling part which endows the organism with whatever predictability its movements have, and if this is not an iron control, but a plastic control under which indeterminacy is not entirely stifled, then it is to be expected that predictions will in principle be possible only within rather tolerant limits" ([1974], p. 393). It should also be noticed that two dimensions of predictability may be discerned in 'plastic control' situations. I shall call them 'degree of predictability$_1$ '(DP_1) and 'degree of predictability$_2$' (DP_2). DP_1 stands for the residual unpredictability which is due to the plasticity of control under which 'indeterminacy is not entirely stifled'. And this residual unpredictability is, according to both Popper and Watkins, inescapable in both the natural and the social sciences. This admission nevertheless does not necessarily render descriptions of actions deductively unexplainable and unpredictable. As Watkins suggests in his [1974], an organism may be viewed as a deep hierarchy of plastic controls. And although at each 'control level' some degree of unpredictability may be encountered, when we come to the prediction of the aggregate effects of the operation of numerous plastic controls, our predictions may turn out strikingly accurate! (Mainly because the minor 'residual' divergencies would tend to cancel out.) DP_2, which I have represented along the vertical dimension on *Schema I*, emerges with the realisation that different situational configurations – all of them generating DP_1 – may plastically control behaviour with various degrees of 'hardness' or 'softness'. That is, in some cases the relevant configurations of mental states ('world 2') and mental constructs ('world 3') may narrowly delimit a behavioural option (that would be an instance of 'hardness' of control) while in other cases the configuration of 'world 2' and 'world 3' objects lead the organism to an extensive variety of alternative behavioural options (that would be an instance of 'softness' of control). It is a consequence of the foregoing analysis that predictive success largely depends on the DP_2 'reading' along the vertical dimension.

[49] The distinction between descriptive and normative rationality theories is important in this connection. A normative rationality theory may be a good guide to successful problem-solving under certain conditions and yet be disastrous as an explanatory theory of problem-solving behaviour. (Cf. above, footnote 24).

[50] Aristotle, *Nichomachean Ethics*, Book III.

[51] Cf. Latsis [1972], [1974], [1976*a*] and [1976*b*].

[52] The above discussion is in many ways an oversimplification. As I mentioned above (see Section 5, footnote 46) any individual organism may be viewed as being subject to a hierarchy of plastic controls. This realisation has the unsettling effect of making the same behavioural sequence appear simultaneously subject to both 'hard' and 'soft' plastic control. Consider an example: A man's aim is to get home in time for supper and he decides it will be quicker to walk than to wait for the bus. So far his aims and

situational analysis may exercise 'hard' plastic control on his behaviour as described. But the analysis, at this hierarchical level, does not fully determine his precise bodily movements. So, his bodily movements may well be subject to 'soft' plastic control.

Fortunately, the complication is not as serious as it may at first appear. Whether we characterise a situational analysis as exercising 'hard' or 'soft' control depends on our initial problem situation. If our starting point is the question, 'Why did X decide to walk home?', then our answer could invoke a situational analysis which exercised 'hard' plastic control. But if our starting point is the question, 'Why did X follow the precise path g on his way home?', then we might find it difficult to supply a similarly 'hard' situational analysis. (This note was added following a suggestion by John Watkins.)

[53] Aristotle, *Nichomachean Ethics*, Book III.

[54] *Ibid.*

[55] Cf. Simon [1976]. But see also Simon's important [1955] and [1956] papers.

[56] Cf. Latsis [1976*a*] and [1976*b*].

[57] For applications to cognitive psychology see Simon [1969].

[58] See also my [1974], Chapters II and VII.

[59] Wiggins made this point forcefully in his [1975]. "Deliberation is still *zetesis*, a search, but it is not primarily a search for means. It is a search for the *best specification*. Till the specification is available there is no room for means. When this specification is reached, means-end deliberation can start, but difficulties which turn up in this means-end deliberation may send me back a finite number of times to the problem of a better or more practicable specification of the end" (Wiggins [1975], p. 38).

[60] Tversky and Kahnemann [1973], p. 20.

[61] See for instance Thibaut and Kelley [1959].

REFERENCES

Alexander, R. D. [1975]: 'The Search for a General Theory of Behaviour', *Behavioural Science* **20**, 77–100.

Aristotle: *Nichomachean Ethics* (Translated by F. H. Peters, Second Edition, 1884), K. Paul Trench, London.

Ashby, W. R. [1952]: *Design for a Brain*, J. Wiley, New York.

Donagan, A. [1970]: 'Comment', in R. Borger and F. Cioffi (eds.), *Explanation in the Behavioural Sciences*, Cambridge University Press, pp. 218–227.

Feigl, H. and Meehl, P. E. [1974]: 'The Determinism-Freedom and Body-Mind Problems', in P. A. Schilpp (ed.), *The Philosophy of Karl Popper*, I, Open Court, La Salle, Ill., pp. 520–559.

Hayek, F. A. [1967]: *Studies in Philosophy, Politics and Economics*, University of Chicago Press.

Hayek, F. A. [1973]: *Law, Legislation and Liberty*, Volume I, *Rules and Order*, Routledge and Kegan Paul, London.

Hempel, C. G. [1965]: *Aspects of Scientific Explanation and Other Essays in the Philosophy of Science*, The Free Press, New York.

Latsis, S. J. [1972]: 'Situational Determinism in Economics', *The British Journal for the Philosophy of Science* **23**, 207–245.

Latsis, S. J. [1974]: 'Situational Determinism in Economics', unpublished Ph.D. Thesis, University of London.

Latsis, S. J. [1976*a*]: 'The Limitations of Single-Exit Models: Reply to Machlup', *The British Journal for the Philosophy of Science* **27**, 51–60.

Latsis, S. J. [1976*b*]: 'A Research Programme in Economics', in S. J. Latsis (ed.), *Method and Appraisal in Economics*, Cambridge University Press, pp. 1–44.

Leijonhufvud, A. [1968]: *On Keynesian Economics and the Economics of Keynes*, Oxford University Press, New York.

Leijonhufvud, A. [1973]: 'Effective Demand Failures', *Swedish Journal of Economics* **75**, 27–48.

Leijonhufvud, A. [1976]: 'Schools, "Revolutions" and Research Programmes in Economic Theory', in S. J. Latsis (ed.), *Method and Appraisal in Economics*, Cambridge University Press, pp. 65–108.

Loasby, B. J. [1976]: *Choice, Complexity and Ignorance*, Cambridge University Press.

Machlup, F. [1946]: 'Marginal Analysis and Empirical Research', *The American Economic Review* **36**, 519–554

Machlup, F. [1952]: *The Economics of Sellers' Competition: Model Analysis of Sellers' Conduct*, Johns Hopkins Press.

Machlup, F. [1967]: 'Theories of the Firm: Marginalist, Behavioural, Managerial', *The American Economic Review* **57**, 1–33.

Machlup, F. [1974]: 'Situational Determinism in Economics', *The British Journal for the Philosophy of Science* **25**, 271–284.

Mises, L. von [1949]: *Human Action; A Treatise on Economics*, Yale University Press, New Haven.

Myrdal, G. [1954]: *The Political Element in the Development of Economic Theory*, Harvard University Press, Cambridge, MA.

Pareto, V. [1899]: *I problemi della sociologia*, Scansano, Tessitori. Reprinted from *Revista italiana di sociologia*, March 1899.

Pareto, V. [1917]: *Traité de sociologie générale*, Payot, Lausanne and Paris.

Parsons, T. [1937]: *The Structure of Social Action*, McGraw Hill, New York and London.

Popper, K. R. [1945]: *The Open Society and Its Enemies*, George Routledge and Sons, London.

Popper, K. R. [1957]: *The Poverty of Historicism*, Routledge and Kegan Paul, London.

Popper, K. R. [1966]: *Of Clouds and Clocks: An Approach to the Problem of Rationality and the Freedom of Man*, Washington University Press, St. Louis.

Popper, K. R. [1967]: 'La rationalité et le statut de principe de rationalité' in E. M. Classen (ed.), *Les fondements philosophiques des systèmes economiques: Textes de Jacques Rueff et essais rédigés en son honneur*, Payot, Paris, pp. 142–150.

Popper, K. R. [1968*a*]: 'On the Theory of the Objective Mind' in L. Gabriel (ed.), *Proceedings of the XIVth International Congress of Philosophy*, Herder, Vienna, pp. 25–53.

Popper, K. R. [1968*b*]: 'Epistemology without a Knowing Subject' in B. van Rootselaar and J. F. Staal (eds.), *Logic, Methodology and Philosophy of Science III*, North-Holland, Amsterdam, pp. 333–373.

Popper, K. R. [1972]: *Objective Knowledge*, Clarendon Press, Oxford.

Popper, K. R. [1974]: 'Watkins on Indeterminism as the Central Problem of My Philosophy', in P. A. Schilpp (ed.), *The Philosophy of Karl Popper*, II, Open Court, La Salle, Ill., pp. 1053–1059.

Rothchild, K. W. [1947]: 'Price Theory and Oligopoly', *Economic Journal* **57**, 299–320.

Simon, H. A. [1955]: 'A Behavioural Model of Rational Choice', *The Quarterly Journal of Economics* **69**, 99–118.

Simon, H. A. [1956]: 'Rational Choice and the Structure of the Environment', *Psychological Review* **63**. Reprinted in H. A. Simon, *Models of Man*, J. Wiley, New York, pp. 261–273.

Simon, H. A. [1969]: *The Sciences of the Artificial*, M.I.T. Press, Cambridge.

Simon, H. A. [1976]: 'From Substantive to Procedural Rationality', in S. J. Latsis (ed.), *Method and Appraisal in Economics*, Cambridge University Press, pp. 129–148.

Thibaut, J. W. and Kelley, H. H. [1959]: *The Social Psychology of Groups*, Wiley, New York.

Tversky, A. and Kahnemann, D. [1973]: 'Judgment under Uncertainty: Heuristic and Biases', paper delivered at the Fourth Conference on Subjective Probability, Utility and Decision Making, in Rome, September 1973. Mimeographed.

Watkins, J. W. N. [1952]: 'Ideal Types and Historical Explanation', *The British Journal for the Philosophy of Science* **3**.

Watkins, J. W. N. [1957]: 'Historical Explanation in the Social Sciences', *The British Journal for the Philosophy of Science* **8**, 104–117.

Watkins, J. W. N. [1970*a*]: 'Imperfect Rationality', in R. Borger and F. Cioffi (eds.), *Explanation in the Behavioural Sciences*, Cambridge University Press, pp. 167–217.

Watkins, J. W. N. [1970*b*]: 'Reply', in R. Borger and F. Cioffi (eds.), *Explanation in the Behavioural Sciences*, Cambridge University Press, pp. 228–230.

Watkins, J. W. N. [1974]: 'The Unity of Popper's Thought', in P. A. Schilpp (ed.), *The Philosophy of Karl Popper*, I, Open Court, La Salle, Ill. pp. 371–412.

Watkins, J. W. N. [1975]: 'Three Views Concerning Human Freedom', in *Royal Institute of Philosophy Lectures*, Volume 8, St. Martin, New York, pp. 200–228.

Wiggins, D. [1975]: 'Deliberation and Practical Reason', paper given to the Aristotelian Society on Monday, 27th October, 1975.

WERNER LEINFELLNER

MARXIAN PARADIGMS VERSUS MICROECONOMIC STRUCTURES

1. INTRODUCTION

Most social scientists and social philosophers today regard their sciences, such as sociology and economics, as mere cognitive, passive disciplines that have the sole purpose of obtaining the best possible knowledge of the past and present states of our society. We may call this well-known view the 'cognitive view'. It seems that only a minority of social scientists and philosophers view their sciences as active, dynamic disciplines with which they can not only invent and construct new societal structures and models, but also feel the obligation to realize models of a new society. Marx was a clear case of the latter type. However, whereas he was quite sure that the social scientist, as a scientist, has to begin historically and cognitively, he was never sure of his role as a realizer of the new. Should he be a mere architect or a revolutionary? An architect, e.g., should, according to a plan, propose better housing via continuous changes of existing housing, or proceed, as Marx did in *Capital*, in a democratic Keynesian manner; a revolutionary might even do this without a plan by appealing to the masses directly. He might appeal for a radical overthrow of the old in favor of something radically new as Marx did in the *Manifesto*. But appealing to the masses cannot be done in a merely rational, scientific manner. One has to conjure up the deeper layers of the human psyche, i.e., the emotional, irrational, and mythical foundations of social life. Marx did exactly that. He conjured up an almost religious paradigm of alienation – and entrenched belief and hope, and mobilized social anxieties, all to achieve his socio-revolutionary goal. In this paper we shall not discuss how this demagogic enterprise has been undertaken by Marx and his political followers, but we shall restrict ourselves solely to showing how and why, under the influence of the paradigm of alienation on the social and economic theories of his time, Marx became a revolutionary. The role of paradigms in revolutionary scientific development and their influence on scientific theories has been discussed widely in recent literature and has also been analyzed by the author in some publications.[1] One of the results of these analyses is that important paradigms are by no means created merely to change sciences; this is only a by-product of their cultural efficiency;

R. S. Cohen and M. W. Wartofsky (eds.), Epistemology, Methodology and the Social Sciences, 153–201.

more importantly, they tend to change society. They resemble far more powerful religious remnants of past cultures and societies that only secretly and randomly influence scientific development, very often simply by being the contrary position to all that the current science teaches us.

Paradigms influence man and his society a great deal, but they are not the only determining factors in the background knowledge of a scientific social theory, especially in social theories. In addition, we have to take into account such inherited factors as personal maxims expressing man's more or less egotistic nature. Among these factors are: strivings such as the maximization of utility (M), philosophical or societal principles (P) of a more or less altruistic ethical nature such as equality (E) and individual freedom (I) and scientific superhypotheses (SH).

For the sake of analyzing the influence of all these theory-external factors of an economic theory, we shall put these factors together here into one set, which we call the agnate set of a theory. This agnate set consists of maxims, principles, paradigms and scientific superhypotheses which do not necessarily belong to the body of a specific theory, but represent the background knowledge of a certain theory at a certain period of time. Most of the principles used here are the same as those discussed recently in collective choice problems by Sen, Bengt Hansson, Leinfellner and earlier by Arrow.[2] Finally, we have to adduce superhypotheses (SH). These are additional, hidden, relatively independent assumptions, measurement theories, e.g., ordinal or interval scaling (SS) of utility, which are *nolens volens* used to understand scientific theories. In this paper we shall show how two different paradigms, together with specific maxims, principles and superhypotheses have led to a twofold interpretation of one and the same fundamental structure K of microeconomics. This structure we will call the 'Ricardo-Smith' or 'Shubik-Shapley' structure of microeconomics. Using the formalism and epitheoretical methodology introduced by the author in former articles and books,[3] we want to demonstrate that the Marxian foundation of microeconomics and macroeconomics is based in part on the same fundamental structure of microeconomics, K, which was introduced and used by Malthus, Ricardo and Smith, but differs from theirs because of the influence of the alienation paradigm. This influence alters the preferentially interpreted value or decision-oriented model of K. We shall do this by a reconstruction and axiomatization of Marx's microeconomic theory as an ordered pair $\langle K, I\rangle$ in accordance with the two-component model of theories.[6] This will consist of a structural kernel K, as defined in section 2, and an intended interpretation I^M, where the superscript M denotes the typical Marxian interpretation of the invariantly given structure K of microeconomics. 'Invariantly given' means that the dynamic structure K represents a pattern of market behavior which has not changed since time immemorial. It has never been rejected, is highly confirmed by market behavior and is considered to be the best-established part, the heart core of microeconomics. It represents the Smith-Ricardo, the classical as well as the neoclassical, game-theoretical, dynamic structure of a market and is based on the works of Shubik and Shapley. The standard or 'capitalistic' interpretation I^C forms again a pair $\langle K, I^C \rangle$ which may be regarded as the received view of microeconomics today. This standard interpretation uses the same kernel K, but a different value-theoretical interpretation I^C.

2. THE FUNDAMENTAL STRUCTURE OF MICROECONOMICS, K

The set-theoretical structure $K = \langle N, X, W; c, f \rangle$ is called a microeconomic fundamental structure, if the following conditions are fulfilled:

A1. : N is a non-empty, finite set, elements n_i, n_j and subsets N_i, N_j belong to N.

A2. : X is a non-empty, finite set, whose elements x_i, x_j belong to X. There are upper bounds for the subsets of X, l_i and l_j, for example, so that the following holds:

$$0 < x_i \leqslant l_i \quad \text{and} \quad 0 < x_j \leqslant l_j.$$

A3. : W is a finite, non-empty set, W_i, W_j are subsets of it.

A4. : c_i, c_j are monotonically increasing functions and C_i, C_j are their corresponding values, so that $C_i = c_i\,(x_i)$ and $C_j = c_j\,(x_j)$ holds.

A5. : P is a non-empty, finite set, P_i and P_j are subsets of P, values of the following monotonically decreasing, twice differentiable functions f:

$$P_i = x_i f(x_i + x_j) \quad \text{and} \quad P_j = x_j f(x_i + x_j).$$

A6. : For x_i, x_j there are the following functions, whose values are the sets

$$W_i, W_j\text{:}\ \ W_i(x_i, x_j) = x_i f(x_i + x_j) - c(x_i) \quad \text{and} \quad Wj(x_i, x_j) = = x_j f(x_i + x_j) - c(x_j).$$

From these axioms of K we may derive the following well-known theorems:

T1. : If x_j is constant, then the first derivative f' is given by:
$dW_i/dx_i = 0$;
and if x_i is constant then f' is given by:
$dW_j/dx_j = 0$.

T2. : There exists a partial differential equation for i and j:
$\delta W_i/\delta x_i dx_i + \delta W_i/\delta x_j dx_j = 0 = dW_i$ and $\delta W_j/\delta x_i dx_i + \delta W_j/\delta x_j dx_j = 0 = dW_j$.

T3. : $W^* (x_i, x_j) + W^*(x_i, x_j) = \max ((x_i + x_j)f(x_i + x_j) - c(x_i) - c(x_j))$.

T4. : $W^{**} = \max_i \min_j (W_i - W_j) = \min_j \max_i (W_i - W_j)$, (Minimax-theorem).

We regard this structure K as an empirical, invariantly given structure of the Shubik-Shapley type,[4] which has been axiomatized by the author elsewhere.[4] This structure K is from a formal point of view a typical classical, effective Boolean structure, or equivalent to a neo-Bernoullian-behavioristic value and decision theory under certainty and no risk. It is a completely strict or orthocomplemented structure. The quasi-ordering of values of ordinal ranking as well as Marx's labor value (rational scale of values) can be formally included into this system K, the logic of which has been analysed elsewhere by the author.[5] Its logic is not quite identical with that of the classical sentential calculus or predicate calculus and it is actually an intuitionistic type of logic. This kind of logic is employed because of the strict, effective sense of the negation used in economic sciences, and because it is important for understanding the dialectic method Marx imposes upon it. This logic can be easily embedded into any classical system of predicate logic, or into a Suppes structure for purposes of measuring extensive quantities, e.g. weights, but it is not equivalent to a non-Boolean structure of the von Neumann-Morgenstern type or with those structures used for interval measurement of value (preference systems under risk and uncertainty). Adding a von Neumann-Morgenstern structure, i.e. a set of axioms, for measuring values with an interval scale changes the structure K to a modular structure, which is to be considered as a natural generalization of a Boolean structure. We will come back to the '*differentia specifica*' between the Marxian microeconomics and the standard or capitalistic microeconomics, for today's standard microeconomics is based on a supply and demand (sub) structure (which is equivalent to interval scaling of values (utility) and/or evaluation and decision-making under risk and uncertainty).

3. BACKGROUND KNOWLEDGE AND THE HERMENEUTIC APPROACH IN THE SOCIAL SCIENCES

The main difference between social sciences and natural sciences lies in the fact that social, economic and political theories are dependent on their specific background knowledge whereas the natural sciences are of a far more independent nature. In this paper we shall try to separate the established 'scientific' kernel of microeconomics from its specific background knowledge and shall demonstrate that the difference between the Marxian and the standard version of microeconomics lies in their completely different background knowledge, which leads to a different interpretation of one and the same microeconomic structure. In our terminology, we have to show how the corresponding agnate sets B^C and B^M, which represent two different kinds of background knowledge, have influenced the microeconomic structure K. The pure theoretical and invariantly given content K of microeconomic theory, as that branch of economics which deals with the particular aspects of a market such as price-cost relationship of producer and consumer and with the producer's and consumer's behavior on a market, has been formalized in a first step by means of the set-theoretical structure K.

In a second step we will obtain two different interpretations of one and the same kernel K; $\langle K, I^C\rangle$ and $\langle K, I^M\rangle$, where the superscript C denotes the standard interpretation, and the superscript M the Marxian interpretation, which are obtained by means of two different agnate sets B^C and B^M. For example, the agnate set of the standard interpretation B^C will consist of standard superhypotheses (SH), standard maxims (M), standard principles (P) and standard paradigms (PA). $B^C = (SH, M, P, PA)$ which differ from those of the 'Marxian' agnate set B^M. All of this will help us to understand how a social scientific theory is influenced in a specific 'cultural' or ideological sense by its agnate set, especially by the paradigms. Thus, K, which defines the formal structure of the kernel, together with the agnate sets (B), constitutes hermeneutic models of a scientific theory. (In our case these models are the Shubik-Shapley standard interpretation and the Marxian one.) The kernel K has been axiomatized by the author for this specific purpose elsewhere.[6] Therefore, the axioms which define the formal structure of the kernel K will find empirical interpretations which are not independent of the agnate set B. Isomorphic mapping of I onto K is guaranteed in the case of the standard interpretation. The components of the agnate set representing the background knowledge play, in this epitheoretical analysis, a similar role, as, for example, the maxims (M) in the neo-Bernoullian interpretation of utility, for example, maximization of utility (*M*) or maximization of group utility (*GM*) in *n*-person game theory. These principles (*P*), as mentioned before, were adduced for the first time in Arrow's conditions for democratic or collective choice and were used widely in Sen's theory of collective choice.[9] Their role was neglected in the philosophy of social science and in social philosophy until Kuhn's discovery of the role of paradigms (*PA*) for scientific theories. Since the Kuhn-Lakatos controversy, philosophy of science has become, on the one hand, more and more sensitive to the theory-external determinants of theoretical knowledge, while on the other hand many of these philosophers have grossly exaggerated the role of paradigms. We will show that paradigms are certainly not the only determinants of scientific development via our hermeneutic interpretation. In such a way one can understand the agnate set (*SH, M, P. PA*) as a presupposition of a systematized hermeneutics of science. It is easy to include matching superhypotheses in the specific framework of scientific theories, but any introduction of nonmatching superhypotheses, e.g., interval scaling of utility (*SC*), would change the ordinal topological character of the structure of *K*. In such a case it would enlarge our theory, leading to more axioms. Adding or eliminating maxims is more difficult since they introduce into and impose upon the theory a normative, prescriptive aspect and strength. Maxims are normative or binding, involuntary incentives, motors of instinctive, inherited patterns of behavior for individuals. The principles are self-chosen or imposed rational, altruistic or ethical obligations for all individuals of a certain period. Finally, the paradigms are condensed prescientific views, *Weltanschauungen*, mythologies or religious convictions of individuals of a certain epoch and ethnic background, which may hinder or support scientific views (theories). Thus, the agnate set *B* seems to be the condensed and crystallized content of a cultural hermeneutics (view) of the sciences of a certain epoch and of ethnic provenance, composed very often of contradictory components.

Since Ebeling,[7] hermeneutics has been a doctrine dealing with all methods of interpretations (hermeneutic) which is contrary to the semantic interpretation which deals with linguistic interpretations alone. According to Gadamer, hermeneutics has to include the totality of man's experience of his world or his society and is present before he approaches the world in a cognitive scientific sense.[8] In the epitheoretical analysis of theories introduced elsewhere by the author,[1] a first systematic approach is made to analyse the background knowledge of a theory.

4. THE AGNATE SET B^C OF THE STANDARD INTERPRETATION OF K

Seen from the hermeneutical point of view we have superhypotheses, maxims principles, paradigms forming the agnate set B^C of a specific hermeneutic position of interpretation. We do not claim here that the agnate set B^C is complete, but we do claim that it is a sufficient precondition for understanding the difference between the standard interpretation of K and the Marxian interpretation. (This again will be used to support the claim that K is an invariant structure.)

4.1 *Superhypotheses*

In microeconomics, which deals with preferences and values, we have to agree about a binding convention with respect to the scale of values. Therefore, we adduce the interval scale (*SC*) of utility as the presupposition for using the structure K within modern economics. Since we need the exact computation of the second derivative f' of our function f, we cannot use an ordinal scale alone. This superhypothesis, if added to the axioms of K, introduces a von Neumann-Morgenstern type of utility theory under risk and uncertainty into our structure. One may, of course, regard *SC* as a structure defined by separate axioms as, for example, in the von Neumann-Morgenstern axiomatization or in the author's general axiomatization of *SC*.[38] The label 'superhypothesis' indicates simply that this presupposition is, in fact, always a scientific and independent one, which may be defined again as a model (an axiomatized set-theoretical structure). This will not be used in the Marxian interpretation, when we shall use Marx's labor theory of value (LThV) instead of it. Since Marx's labor theory of value introduces a ratio scale, it is stronger than *SC*. Therefore, LTh $\rightarrow$ *SC* but not vice versa. The introduction of *SC* has two far-reaching effects on K: (1) Formally it induces modularity into K. Thus K is changed from a classical Boolean to a non-classical modular structure, e.g. by the axioms of group 3 of von Neumann-Morgenstern's utility theory. This has been analysed elsewhere by Leinfellner and Booth.[5] (2) In a theoretical sense, it enables us to represent mathematically and isomorphically the consumers' and producers' or the buyers' and sellers' preference behavior under full risk and uncertainty on a linear vector space of values – the most common probabilistic value space of all social sciences. Interestingly, this is not possible in (K & ThV) alone.

4.2 *Maxims of the Standard (Capitalistic) Interpretation*

The (neo-utilitarian) maxim of maximization of individual utility (M) is often called 'individual rationality': $v(i) \geqslant 0$. This is certainly one of the most famous hidden assumptions of modern microeconomics. Note that the Marxian superhypothesis, the law of maximization of capitalistic profit (LM), can actually be obtained from the more general maximization of utility M. Therefore: $M \rightarrow LM$. Another maxim is group rationality (GM) in the form $v(N_1 \cup N_2) \geqslant v(N_1) + v(N_2)$, where the $(N_i \cup N_j)$ are groups or classes in our society. It clearly symbolizes the striving of the individual to unite with others to form classes. One may call it the class formation maxim; it expresses value-theoretically the inherited tendency of man to form hordes and tribes as well as classes and unions. Any individual has only one reason to join a group, if he is better off in it than out of it. From a formal point of view, the above-mentioned group maxim induces a superadditivity of probability into our structure K.

4.3 *Principles of the Standard Interpretation*

Whereas the superhypotheses always express well-established scientific assumptions, e.g. about mass measurement or about the utility (value) scale which may be used in microeconomics, and the maxims express the individual's egotistic strivings, the principles denote general (altruistic) perspectives, social goals or ethical norms of a certain period. It seems that the maxims and principles are objectifications or formulations, splinters from the paradigm, which have often undergone codifications in constitutions, for example. It has been said very often that at least some leading principles of a free market in a free society play an important 'democratic' role, such as equality (E), which means here a functional equity in which one member of the society may replace the other and vice versa.

If we have an unrestricted domain of preferences (U), freedom is independence of irrelevant alternatives (I), nondictatorship $\bar{D}$ and Pareto optimality (P). It is well known that for collective democratic decision-making, as Arrow has shown, from the above introduced 'agnate' set of principles, which he called "conditions" of democracy: $(U, I, P, \bar{D})$, where $\bar{D}$ means dictator-free, plus the ordering axioms for ranking and collective choice, the well-known Arrow paradox follows:

$$\{(U, P, I, \bar{D}) \,\&\, (0)\} \rightarrow D$$

which is equivalent to the contradiction: $\bar{D} \to D$. This result has been generalized later by Sen and Schwartz[9] in such a way that it covers all kinds of social and economic decision procedures. Therefore, our market model is contaminated from the beginning by Arrow's paradox. It is not Arrow-immune, since it is a typical decision structure under uncertainty and risk. As a consequence we have to give up the claim that our structure K in any decision-oriented form can be based on the 'Arrow-principles' $U, I, P, \bar{D}$. This is true for either the Marxian or the standard interpretation, with the difference that in the former, dictatorship is explicitly permitted.

4.4 *The Standard or the 'Capitalistic' Paradigm*

It is not too difficult to describe a standard paradigm of market behavior, because it is the original source for the game-theoretical foundation of microeconomics. Huizinga proposed to redefine man as a playful being (*homo ludens*), instead of a 'social being'.[10] It is more neutral than, and furthermore it is not so paramount as the dominant and ethical cultural weight of the alienation paradigm for Marxian microeconomics. The market, as the partial social stage on which economic actors – firms, households, consumers, laborers, producers – meet and make their economic decisions, is in reality the prototype of a game or decision framework consisting of conventional rules for actions and decisions. The decisions are dependent primarily on preferences, demands and evaluations. Prices (P_i), wages (C_i) and profits (W_i) are monetary values, and the demand and supply, as well as the cost functions are utility (value) functions. There is a genuine uncertainty and risk in the market, which the participants are willing to calculate and to share. All of this is submerged in our structure K but without being a specific game such as exploitation. Yet, since Weber, Troeltsch and Cunningham,[11] we may ask the question: Is economic enterprise in reality not dependent more on the deeper entrenched Protestant and Calvinist religious belief? Did not the Protestant ethic, 'God with us,' create the spirit of capitalism as Weber depicted it, and never the spirit of gambling? But history tells us that capitalism developed again and again in the course of very divergent social and economic developments in different nations, without any contact at all with Protestant or Calvinist religious ideas, e.g. such as in the capitalistic period of Rome. It seems as though that (as Brentano and Tawney[12] have already outlined it) in fact gambling, within a framework of rules fixed by a social contract, is a more general presupposition or paradigm for explaining capitalistic economic development than the Protestant or Calvinist paradigm. Certainly the Protestant ethic was only one factor which contributed to the rise of the free capitalistic economic system and the free market. It is not difficult to see that, at a certain stage of cultural development, man begins to understand and believe in the efficacy of rational conflict resolution offered by 'games' as social decision frameworks. This analytical tool enables him to perform single, dual and collective decisions openly and systematically. The method is fair since everybody knows in advance what he may expect to win or lose in the course of the game; this method seems to be free because the participants have the greatest amount of self-chosen possible freedom, e.g. they may always choose from a given variety of moves, actions.

The 'game' is evolutionary because it may be easily changed without deep-going revolutionary ruptures by just changing the rules, for example from a discriminative to more socially tolerable outcomes. Finally 'games' open up avenues of success for everyone who is skilled enough to discover winning strategies or is willing to learn from past experiences.

This paradigm of a 'game' is certainly of a more practical empirical origin than the democratic and liberal ideas of unrestricted freedom which, in fact, stem from religious equality among individuals, etc., because a 'game' is simply based on the rational and practical assumption that actions and decisions need a general 'restrictive' consensus, a social contract concerning the best possible rules and decision procedures for making decisions, either for the society or for the individual. The predecision, of course, to partake in such a game should always rest on the individual. He has to choose freely according to the outcomes and expected outcomes of different possible games (in different societies) which one he may select as his own. Whereas equality in a Platonistic and religious sense regards all men principally as equal, the game-theoretical paradigm regards men as being relatively equal, i.e. he can choose only amongst certain generally accepted and available 'free' moves and 'free' actions defined by the rules of the specific game. He possesses only a certain leeway of freedom, never absolute freedom of 'unrestricted' choice, as Arrow assumed. We are now in a position to reconsider Weber's ideas and agree with Weber that the Protestant liberation of man from his religious anxiety and slavery in the sixteenth and seventeenth century was the occasion when he opened his eyes to the earthly possibilities of the economic game. Not only the nobility could participate, but everyone capable of learning the rules of the 'game' could do so. In such a sense, 'capitalism for all' was certainly the result of man's religious liberation. We agree now with Weber's argument that according to Protestant and Calvinist religious convictions, God has put the resources of his created world at the disposal of man as a this-worldly reward for the elected or skilled ones. Furthermore, the forces of predestination are always on the side of the elect, the winner in the 'game'. Besides all the pragmatical-practical advantages, there is one tremendous disadvantage of the game-paradigm: it loosens man's social responsibility. Unluckily this trend has been enforced by the Calvinist doctrine of predestination, according to which man's world of culture and the social institutions of his economic life, which constitute the socio-economic game, are at best irrelevant to his salvation. Since salvation is predestined, man is free to act on the social stage, as well as in the market, as he wishes to act. He is restricted only by self-chosen rules and/or democratically self-imposed order. Consequently, when economics later became a normative science, these Calvinist presuppositions crystallized as rules about how to make optimal decisions, i.e. the business man could determine the kind of game and had only to face and calculate the risk and the uncertainty attached to any endeavor and cash the profits when he succeeded. We do not believe that this consequence of 'liberation' is a typical Protestant one, but we assume that only its consequences unintentionally disturbed the balance between the egotistic and altruistic social strivings on which societies are based. This equilibrium – characteristic of ideal societies – shifted to the egotistic side, and man followed eagerly his inborn maxims of maximization of his own utility and greed. Within the variety of 'different games' and different admissible rules of an economic game, such as the market, one may say that whenever in history, whether by chance or by developmental factors, man's altruistic social incentives were weakened, such a 'dissocializing' shift occurred. Dissocialization by playing a game is due to the fact that the goal and motor of any game is simply that the individual should maximize his personal utility. The individual will feel himself restricted by impersonal rules, i.e. by the permitted moves and by his abilities to win, but he will not likely ask himself: should we not improve the rules, is it good for all to maximize egotistically their utility? Everyone will take his chances and win if it is possible for him to do it and will regard altruistic behavior as something he has to avoid, since it leads to personal losses for him. But games are not immoral *per se*; often dissocialization is followed by 'socialization'. This was Marx's belief. We need, of course, to explain more the whole history of capitalism, which we defined as a weakening of man's altruistic and social tendencies by considering, for example, the technical and industrial revolution, the rise of mass societies, etc. For our epitheoretical analysis, it is important to emphasize that the maxims which we are all very willing to obey became, in fact, too important. This is the result of the paradigm of the 'game' for Western microeconomics. Thus we face a real ethical dilemma: On the one hand, if we maintain the maxims of maximization of our own utility, we become more and more socially irresponsible and unjust. On the other hand, without the maxims 4.2, the whole micro- and macroeconomic system would not work. Any practical weakening of the maxims 4.2 in recent history, either by political Marxist doctrines or by nationalization, weakened at the same time the whole present system of economics and economic production. From this point, we see clearly that recent ethical foundations of economics or welfare economics are in reality a counter-move against dissocialization. Nevertheless it follows the preassigned lines of emphasizing the Puritan ideals of fair game by understanding the rationale of the economic game. Once the game-theoretical and decision-theoretical character of microeconomics in the midst of the twentieth century was 'theoretically' established, the next step was either to improve the rational character of games and their rules with ethical standards, e.g. fairness, readiness to compromise (minimaxing), etc., or to regard justice as a distributive ethical principle for a fair division game, etc. Harsanyi's welfare economics, Arrow's and Sen's foundations of collective choice on disguised ethical principles, and Rawls' distributive justice are examples of reintroducing ethical principles into capitalistic microeconomics.[13] This amounts to an ethical correction and improvement of an already existing game.

5. THE AGNATE SET B^M OF THE MARXIAN INTERPRETATION OF K

5.1 *Superhypotheses*

We regard the three very well-known 'laws' of capitalistic economy in Marx's writings as typical Marxian superhypotheses: $SH1$: The ever decreasing surplus value of labor. $SH2$: The ever increasing concentration and centralization of capital (monopolization). $SH3$: Maximization of profit (ML). Another one is the famous labor theory of value of (LThV) or quantizing (scaling) of values in proportion to standardized labor time. (Note that $M \rightarrow LM$ and LThV $\rightarrow SC$ hold in a simple commodity production economy.) Some people assume the law of class struggle as some kind of superhypothesis. In this paper we shall show that, with exception of the labor theory of values, all Marxian superhypotheses or laws of capitalistic development are indeed theorems which follow from our axiomatized structure K, i.e., from the game.

5.2 *Marxian Maxims*

Marx introduces maximization of utility (*M*) ambiguously and indirectly. Firstly he assumes that maximization of utility *M* holds only for the capitalist, not for the laborer, whenever the capitalist tries to maximize his profits (*LM*) by exploiting the worker. Secondly, at the moment the laborers counter the capitalist and unite for class struggle, they follow group rationality (*GM*). Historically seen, Marx assumes the paradoxical behavior that the laborer awakens suddenly from his slumber by facing exploitation. Then he reacts by uniting and learns to maximize his individual utility (wages), but this maximization of utility (wages) will abruptly cease, once the laborer belongs to a disalienated society, e.g. an ideal communistic one. It is clear that group rationality (*GM*) holds since it follows from the rationale of the Marxian motto 'Workers unite'.

5.3 *Marxian Principles*

The three famous principles of dialectical materialism (*PDM*): unity and strife of opposites, quality-quantity changes, and negation, belong more to the post-marxian development, but we will show how they can be derived from the alienation paradigm (*AP*). ($AP \rightarrow PDM$). In his work, Marx additionally uses, from time to time, the principle of market equilibrium (*EQ*) and Pareto optimality (*P*).

5.4 *The Marxian Paradigm of Alienation*

This section belongs in reality to a philosophy of labor. It deals with the dilemma that the world in which we live is the product of human labor, yet we feel uneasy in this world, alienated.

5.4.1 *The Historical Background of the Alienation Models.* The author has delineated in detail elsewhere the historical sources of the alienation paradigm.[3] It can be traced back to religious, mythical models of alienation which are as old as Western culture and philosophy. We find it in two versions from the beginning: the realization paradigm of alienation and the psychic alienation paradigm, both of which we shall describe in models, as has been done extensively elsewhere.[3] In both versions a person (god, artist, laborer) changes under the influence of creative activity from a state of harmonious identity and unity with his work and product to a state of lost identity and disunity (alienation). From this state he finally finds his way back to a restoration of his former unity at a new and different level (disalienation in a utopian society).

By a model, we understand any linguistically reliable (isomorphic) representation of structures (regardless of whether we deal with empirical, social, inner religious, or psychic structures). We may place the two fundamental models of alienation as protomodels for labor at the beginning of Western philosophy, the realizing model and the psychic model. From the first, the realizing model, we may derive the ontotheological, the artistic, the Hegelian-Marxian model of alienated labor, and most important, finally the dialectic method and Marx's labor theory of values. The other perennial line of the psychic model begins with the religious mythical models of alienation, set forth in the gnostic and mystical models, and finally culminates in the secularized model of Romanticism (including the psychoanalytical one). Marx has used the Janus face of especially the psychic model as have also Heidegger and Marcuse, as a castigating critique of inner estrangement and frustration of modern man in his alienated (social) world. Both lines of conflicting thoughts stretch over a period of about 2000 years of civilization; they represent a continuous and (until now) unsolved problem of social and inner uneasiness. The conflict is always a consequence of creation and creative work of something completely new if it is done by God or gods, as in the early religious versions, or is done by creative artists, as in the secularized labor-version, or finally, is done by the productive technician or the laborer as in the Feuerbach-Marx materialization version, which is simply the Hegelian version turned upside down. The last act of a worldwide drama (which has been discussed by the Club of Rome) may begin immediately with the total and mondial alienation of man in his world as a consequence of man's worldwide social, economical, and technical, 'creative' changes leading to a catastrophic deterioration of his ecosphere and his planet earth.

5.4.2 *The Religious Model of Alienation*. One may find this model in the version of the prodigal son in the Bible, in the apocryphal acts of Thomas as 'Hymn of the Soul', or as a gnostic version in Bardesanes.[14] It is a mythical and mystical narration of the conflict evoked by the 'European idol' of activity and creativity from which even God is not excluded. The narration is as follows: The son of God, who chooses, out of his free will, the activity of incarnation into the world and experiences the otherness, the misery of the world, becomes alienated. It is his 'free will for activity' which is the source of alienation. But by a kind of learning process he returns and is reconciled and disalienated by the (social) fact of being evaluated higher than those sons of God who stayed all the while with God in inactivity and rest. This is exactly the passage from which the term 'alienation' comes, which found its way via Luther and Hegel to Marx. Strangely enough, one may already recognize the triadic, exclusive steps of a dialectical development: (1) unity-thesis; (2) disunity-otherness-alienation-antithesis; (3) disalienation-reconciliation-synthesis or return. Since we are dealing with the inner or psychological model of alienation, there should be no doubt that the three states are inner experiences. The stimulus here for being alienated from God or from the primordial unity lies exclusively in the sudden awakening of individual activity and the creative actions in a part (son) of God. Estrangement can only occur after it has been actually experienced and if the alienated state is completely new. Only after experiencing alienation can there be a turning point towards disalienation. Disalienation is here, in the psychic-religious paradigm, practically an inner reconciliatory step in the mind of the alienated being, i.e. this reconciliation is achieved only in and within man's consciousness. If we understand the message of this myth correctly, it means that when social alienation by labor occurs, disalienation can be achieved only by culture, education (as recently done by Japanese management of labor), by enlightenment, philosophy, or complete information about labor and so on, and not by a total and absolutely equal material (re-) distribution of all goods alone. This model or paradigm appears in the Osiris-Isis myth, the Sinbad-Odysseus-myth, and in *Faust*. It has been used by Jung, by Kerényi and by phenomenology of religion as the description of the inner conflict occurring in those possessing god-like creative and innovative activity. This paradigm has been described as the archetypal symbol of the West's worship of activity in many works of art.

5.4.3 *The Romantic Model*. This model may be obtained from the religious model by substituting nature for God as the primordial original unity. Rousseau's 'back to nature' means avoiding alienation by all means, but only after the disadvantage of alienation has been experienced by man. Man needs to go through a learning process of alienation. This kind of secularization expresses a trend, which will later be concluded by Feuerbach's and Marx's 'materialization' of the realization model of alienation which we will discuss briefly in the following pages.

5.4.4 *The Ontotheological Model: The Protomodel of Idealistic Philosophy*. The ontotheological model has to do with the parable or the mythologema, that God has created the world as something completely different from himself and is therefore alienated by his own, now 'alien' creation. He has to reconcile himself with his creation, for example, by introducing salvation of the world by his son. The ontotheological model may be the most comprehensive paradigm of idealistic philosophy. The first phase of this model is initiated exclusively by a process, realization of work; the second phase of disalienation remains as an inner change or return, analogous to the religious model. This paradigm is present in nearly all religious and mythical accounts of the creation of the world, e.g. Plato's, the Neoplatonist's, the gnostic's and, of course, the Judaeo-Christian, as well as the Augustinian explanation of the creation of the world. Via medieval mysticism, Böhme's *Aurora*, this model enters German idealistic philosophy and is actually the basic skeleton of Schelling's and Hegel's philosophical system. In all these systems, we find that alienation is the first phase in the active creation, disalienation the second phase, whereas the third phase is an inner reconciliation in the mind of man, e.g. the assumption of Christ, the ekstasis of Plotinus, the salvation of man. In Hegel's system, the awakening of self-consciousness means disalienation in the widest sense. Hegel's system, in which the idea (god) in the state of being-in-itself (thesis) is dismissed to its otherness, nature, the antithesis, or the being-for-itself, from which the idea ascends via experience, science, culture, philosophy, and religion to self-consciousness or to the being-in-itself and for-itself (reconciliation or synthesis or disalienation), is nothing less than a gigantic extrapolation of the ontotheological model of alienation in terms of a philosophical system. Hegel's tendency towards methodological abstraction finally results in his dialectical method (or 'logic' as he calls it), in which the model of alienation finally disappears in the innocuous and formal triadic steps of his logic. But whenever Hegel is required to furnish practical interpretations of his 'logical dialectic', he chooses the alienation or self-alienation model. Historically seen, the situation is just the reverse: the paradigm of alienation creates the famous dialectical method. Dialectic, in the Hegelian and in the materialistic Marxian sense is, therefore, an offspring of alienation itself. We will prove that in the next section. To conclude this historical excursus, we want to describe in a few sentences the artistic model of alienation (which played an important role in Plato's demiurgos concept, in Schelling's replacement of god by 'god as artist', and in the Romantic concept of the artist). What was earlier described by Marx as objectification or materialization of human energy, of life or simply of life in the product of creation by labor, is symbolized in H. Hesse's *Journey to the East* as two figures, one symbolizing the artist as an individual, the other the artist as work. The first figure declines steadily, whereas the second flourishes proportionally. With his alienation model of labor, Marx describes the situation as follows: "Any good has value only because labor in the abstract has been embodied or materialized in it."

5.4.5 *The Alienation Model and Dialectical Method*. As we have pointed out earlier, Hegel's philosophy, as well as Marx's dialectical materialism, begins with the famous dialectical method, and the model of alienation is used whenever appropriate interpretations of the dialectical method are sought. For Marx the most important interpretation was the interpretation of creation by labor and the process of labor in the 'master and slave' chapter in Hegel's *Phenomenology of Mind*, an earthly version of the god-son parable. This interpretation has not only been adopted fully by Marx in his alienation model of labor, but indicates the secularization of creation to labor. According to Marx, materialism's main thesis is "being before consciousness" (*Sein vor Bewusstsein*). He had to put Hegel's idealistic system, in which consciousness precedes being, upside down. Feuerbach's and Marx's materialization of the realizing or ontotheological model of alienation is the exact turning point for the conversion of an idealistic dialectic to the materialistic one. After this conversion, the materialistically interpreted alienation model of realization is the model of alienated labor and of alienated technical production. But there is another, more hidden, transformation of alienation, namely into the dialectical process. As used by Marx, it begins with the formation of the opposite in the course of any physical change, motion, or process, and is exactly the formation of something completely different, completely alien to the primordial unity. We will express this formally by the model of a triadic, dialectical structure by defining the formations of strict complementary sets, e.g., $S_2 = CS_1$, which is completely different from the original set S_1 or $S_1 \neq CS_1$ or $S_1 \cap CS_1 = \emptyset$. The term: the 'opposite' or the 'materialized' contradiction is brought directly from the realization model of alienation; likewise the next step of the synthesis (reconciliation) of the opposites S_1 and CS_1 leads to a union: $S_1 \cup S_2 = S_3$, $S_2 =_{\text{df}} CS_1$ or the first synthetic unity set: S_3. This set-theoretical model is the first formal reconstruction of the dialectical method or the dialectical process and should (1) demonstrate the provenance of the dialectical method from the alienation model of realization and (2) explain formally the three basic laws of dialectical materialism, e.g., the law of the unity and strife of opposites which initiates any process, according to Marx. In dialectical materialism, it leads to the opposite state or the state of total alienation (complementary state). Thus 'formation of opposites' and 'formations of the contradictions' are only circumscriptions and parallel expressions for being alienated. The second law of sudden change of quality into quantity and vice versa guarantees the complete isolation and separation of both states. This second law is illustrated by Marx-Engel's famous example of water-and-ice. The final law of the negation of negation avoids or prohibits cycles or loops in any processional development. Whereas the classical law of double negation leads e.g. in the case of CCS_1 to S_1, i.e. $CCS_1 = S_1$, we have to negate in any dialectical process in a different semantic way. The first negation leads from CS_1 to S_2 and the only second dialectical negation permitted leads from the synthesis S_3 to S_4 since $CS_3 = S_4$. We have to remember that effective or intuitionistic restrictions on the negation are similar. For example, the negation of 70°F may be 80°F and the double negation non-non (80°F) may not yield in all cases the first 70°F, but, e.g., 60°F. This does not mean that classical negation: $CCS_1 = S_1$ does not hold in non-dialectical formal Marxist logic, but only that this form of negation cannot explain the dialectical process. It may be interesting to define such a dialectical structure which exhausts a given universe U by the formation of complementary dialectical sets. The set-theoretical model shows clearly, on the one hand, that we are unable to derive the Marxian model of alienated labor or the alienated labor theory from the dialectical method alone. Therefore, the dialectical method follows from the alienation model, but not vice versa. On the other hand, the model explains some methodological features, e.g., the procedural synthesis of a complemented universe of discourse U of the dialectical process in Hegel's logic.

5.4.6 *The Dialectic Structure of U.* Let $(D, U: C)$ be a dialectic structure, if and only if the following conditions are fulfilled.

1. D be a non-empty set of subsets $S_1, S_2 \ldots, S_n$ of U, where $D \subset U$.
2. If the S_i are subsets of D, then the following triadic dialectical structure where $i = 1, 2, \ldots, n$ is given:

2.1 $S_1 \neq CS_1$;
2.2 $CS_1 =_{\text{df}} S_2$ or $S_1 \cap CS_1 = \emptyset$;
2.3 $S_1 \cup CS_1 = S_1 \cup S_2 =_{\text{df}} S_3$;
2.4 $S_3 \neq CS_3$ or $S_3 \cap CS_3 = \emptyset$;
2.5 $CS_3 =_{\text{df}} S_4$;

2.6 $S_3 \cup CS_3 = S_3 \cup S_4 =_{df} S_5$;

2.7 $0 \subset S_n \subseteq U$, where the S_i, $i = 1, 2, \ldots, n$, $S_i \subset D$, $D \subseteq U$, are the only admissible dialectical synthetic sets, which form the universal set U.

In this set-theoretical structure, the set D is a triadically ordered, dialectical substructure of U, where the uneven subsets are the 'theses' and the even subsets the 'antitheses', e.g., S_1 is the thesis and S_2 the antithesis. The definitional equivalence '$=_{df}$' indicates that the antithesis is a specific and separated state or entity. Each synthesis is again a new thesis in the endless spiral of the dialectical structure or process. Complementation C represents the function of effective negation and is always the formation of a completely new and alien set. The complexity of dialectical negation consisting traditionally of 'negare', 'elevare', 'conservare', and 'conciliare' can be explained by: (1) the typical character of alienation expresses the 'negare' function of the dialectical negation and (2) the synthesis, which is represented by the '$\cup$' (or union operation between sets), symbolizes the function of the traditional 'conciliare' of the dialectic and imposes (3) on the synthetic universe U a hierarchical dialectical structure, which symbolizes the function of 'elevare'. We indicate by 'synthetic universe' simply the fact that the dialectic movement and the dialectical process creates the hierarchical dialectic order by means of the set-theoretical operations of our model. Finally, (4) the typical function of 'conservare' of the Hegelian and materialistic dialectic hierarchy is imposed on the sets S of our universe of discourse U. We see very clearly how the model of alienation works as the hidden paradigm, singling out sets or concepts according to the position-alienation-disalienation or thesis-antithesis-synthesis operation. This is exactly the reason why Hegel regarded the disjunctive syllogism as the methodological skeleton of his dialectical method.

Any empirical interpretation of the dialectical procedural structure, e.g., motion or change of systems, uses the fact that the formation of the complementary set or state must be a temporal process. S_1 and S_2 form an overlapping state within a certain period of time as long as $S_1 \neq S_2$ or $S_1 \neq CS_1$ or $S_1 \cap CS_1 = \emptyset$ is not yet fulfilled. Statistical interpretations of these transitional and temporally 'overlapped' stages have been suggested, for example, by Lenin in his commentary on Hegel's *Logic*. Grofman's and Hyman's[15] statistical interpretation of the 'or' would explain this 'overlap' better than Lenin's analysis.

This model of dialectical, triadic structures (1) explains the dialectical method of Hegel and Marx in a formal sense, if we replace the sets by class concepts having the same extension; (2) demonstrates very clearly the origin of the traditional properties of the dialectical negation such as 'negare, conciliare, elevare and conservare'. Thus the dialectical method stems from the paradigmatic concept of alienation or from the alienation model of realization.

6. THE MUTUAL PROJECTION OF THE TWO MODELS OF ALIENATION OR THE UNEASINESS OF ALIENATION IN SOCIETY

It is Marx's original invention to use both models of alienation in the following sense: he projects the realization model of alienation into the psychic one, which we may call in modern psychoanalytical terms 'introversion':

$$AM^R \gg AM^P$$

This means, he starts with an objectively given, concrete situation of estrangement by labor, AM^R,[16] which is changed into a subjectively experienced alienation in the consciousness of man, AM^P. After this projection, $AM^R \gg AM^P$, there follows a counterprojection, extraversion:

$$AM^P \gg AM^R$$

This enforces, by a quasi-feedback, the immiseration of the workers, their frustration and their social uneasiness. This scheme of projection and counterprojection developed into the most formidable, effective criticism, and merciless whip of the capitalistic society. In this scheme (Marx called it an analysis of alienation) it makes no difference where one begins. For example, in *The Economic and Philosophical Manuscripts*, he begins with the objective, concrete alienation model of labor and of property, and proceeds to the alienation of capital, trade, competition, and money. He expresses this very clearly when he writes:

> Just as we have derived the concept of private property from the concept of estranged, alienated labor by analysis, so we can develop every category of political economy with the help of these two factors; and we shall find again in each category, e.g., trade, competition, capital, money, only a definite and developed expression of these first elements.[17]

How does this schematic analysis work, i.e. how is society alienated? In the case of alienated labor, Marx goes back to the alienation model of realization, which he took from Hegel's *Phenomenology*, but not without having turned it upside down, i.e. cleansed the *Phenomenology* with Feuerbach-Marx materialism. It is worth noting that one can never fully understand Marx's social philosophy without understanding the paradigm of alienation, but once trapped in this paradigm, one will never find one's way out of Marx's philosophy. Let us regard once more in detail the projection, introversion $AM^R \gg AM^P$ in Marx's work.

After having materialized and secularized the Hegelian realization model of alienation, the physical process of labor is alienated. Marx starts with labor as realization. Firstly, "Labor's realization is its objectification";[18] secondly, according to the traditional realization model of alienation, any realization has to go through the phase of alienation, provided something completely new is produced or created. Realization is, according to Proudhon and Hegel, exactly the appropriation of the world (or of the raw materials) by means of the value-increasing labor. Hence, the appropriation of goods by capitalistic production is the creation of private property, and if the goods produced do not belong to the laborer, they create estrangement. Thirdly, the projection (introversion) $AM^R \gg AM^P$ begins. That is the reason why, for Marx, creation of private property is not something external.[19] Marx mentions Smith who "no longer looked upon private property as a mere condition external to man." The "subjective essence of private property ... is labor."[20] Thus alienation now becomes an inner disturbing factor of the human psyche and mind, it is internal experience (AM^P). This experience is caused by a certain form of labor which alienates man internally. In the next following act of counterprojection (extraversion) the product of labor will become a twice alienated object. (It becomes objectively alienated private property, "arising as a result of estranged labor, in its relation to truly human and social property."[21]) It is exactly the psychological frustration of anonymous mass-production in our technical modern society which Marx begins to describe. "What, then, constitutes the alienation of labor?" we may ask with Marx. (1) We get the first answer, describing the estrangement of labor (AM^R) as: "The relation of the worker to the product of labor as an alien object exercising power over him."[22] "This relation is at the same time the relation to the sensuous external world, to the objects of nature as an alien world inimically opposed to him."[23]

(2) Secondly, Marx describes the projection $AM^R \gg AM^P$ as alienation, as an internal inner experience.

This relation is the relation of the worker to his own activity as an alien activity not belonging to him; it is activity as suffering, strength as weakness, begetting as emasculating, the worker's own physical and mental energy, his personal life indeed, what is life but activity? – as an activity which is turned against him, independent of him and not belonging to him. Here we have self-estrangement, as previously we had the estrangement of the thing.[24]

In (1) we find alienation as depicted in the alienation model of realization (AM^R) as a fact; in (2) Marx describes the already performed projection ($AM^R \gg AM^P$) and in (3) we will read about the counterprojection. Thirdly:

If the product of labor is alien to me, if it confronts me as an alien power, to whom, then, does it belong? ... The alien being, to whom labor and the product of labor belongs, in whose service labor is done and for whose benefit the product of labor is provided, can only be man himself ... Through estranged, alienated labor, then, the worker produces the relationship to this labor of a man alien to labor and standing outside it. The relationship of the worker to labor creates the relation to it of the capitalist (or whatever one chooses to call the master of labor). Private property [of the capitalist] is thus the product, the result, the necessary consequence, of alienated labor, of the external relation of the worker to nature and to himself.[25]

Marx's early writings throw new light upon the origins of Marxism. Marxism – not communism – culminates in nothing other than a demand for an ethical foundation of creative social labor. Estrangement of labor should be avoided, which comes inevitably from the realization model of alienation. Alienation is a self-enforcing circuit: alienation of the laborer himself is produced by a projection or introversion of estrangement into personal alienation and is an inner self-alienated state of mind and psyche. It can again be counterprojected, 'reinforced' into the external societal and economic environment. It is a deadly disease of modern society and its mechanization of labor and is especially an ethical conflict of modern mass societies. Thus, alienation serves easily as a negative social critique of any already alienated society. In this sense the alienation doctrine of the *Manifesto* has been and will be used (and misused) as the most destructive critique against all kinds of societies where 'laborless' owned property exists. But it is, in reality, far more, for it exposes the necessity for an ethical and social solution to the most important problem of modern society: How to cope with the gigantic power of technical production and creative labor, without falling into the trap of a partially or a totally alienated society. It is the search for an ethical solution to the problem of social labor and productivity, which are the only 'creators' of values, goods, property. The just redistribution of wages and income is only a consequence if one has adopted Marx's view. Generally, it is the ethical foundation of economy which is in the center of Marx's philosophy. But the conclusion for Marx in his earlier writings and in the *Manifesto* is a radical one. All categories of economic and social life in our mass society are contaminated, estranged and alienated;[27] therefore, private property, for example, the means of production, wages, capital are all examples of alienation. Hence, either alientation has to be abolished via a revolution or it has to be positively annulled. Annulment is not disalienation, but refers rather to a social prophylaxis, for the purpose of avoiding alienation in any future human society. It was Marx's early scientific proposal, namely that social sciences, and especially economics have to take care that future labor and production would never become contaminated with estrangement and alienation. Positive annulment is an ethical foundation of economics, is "appropriation of human life", "return of man from religion, family, state, etc. to his human, i.e., social existence".[27] Positive annulment of alienation is Marx's fascinating positive message. The communism of the early Marx, is, therefore, a humanism. Alienation and estrangement, in the play of projection and counterprojection, is henceforth for Marx a merciless, propagandistic critique and whip of any capitalistic society. All the followers of Marx up to and including Marcuse and the existentialist Marxists should be aware of the fact that it is very easy to use this whip, but incomparably more difficult to construct an alienation-free model of society and economics. However, this is exactly what Marx tried to do in *Capital*.

7. THE LABOR VALUE THEORY OF *CAPITAL*

Marx himself wanted to draw the consequences in his new economic theory of *Capital*. The economic system must be changed, the estrangement resulting from the possession of private property as capital or dead labor must be eliminated.[28] As a consequence, Marx had to begin with a new foundation of economics and with a theory of society based on an ethical and social concept which he found in his labor theory of value, a foundation which would not lead his theory of society into the trap of estrangement and alienation.

7.1 *Marx's Value Concept*

For Marx economics is a science of values plus a 'preventive ethical' foundation of labor. His main principles are:

7.1.1 any value (i.e. any increase of the value of goods and commodities) has to be equivalent to an increase in quality (use value = *VU*) or quantity of goods which again has to be equivalent to the value or amount of labor (*LV*) required to produce them: $VU = LV$. But how can we establish such an equivalence? Raw products are only changed by labor into more valuable goods whose values-in-use are increased. Thus, instead of the creation of private property, i.e. of goods and commodities, the parallel creation of values is analysed.

7.1.2 Labor and the amount (of time) of labor put into the creation of values is nothing other than life, or fractions of life-time, i.e., the energy or life of the individual workers. This is Marx's basic ethico-social dogma.

7.1.3 There is no value creation or change other than that which is labor produced. By following 7.1.1–7.1.3 we will see immediately how Marx is setting a collision course with the underlying Smith-Ricardo structure of microeconomics. This conflict will amount to a titanic struggle of the alienation paradigm versus neo-Bernoullian microeconomics. Consequently all Marxist economists are troubled by and dispute again the utilitarian preference (utility) basis of modern microeconomics developed by Walras, Jevons, Menger, von Neumann-Morgenstern, *et al.*

It is interesting that Marx, at the beginning of *Capital*, comes very close to Locke's and Smith's utilitarian foundations of values, when he begins with the natural worth as use value or value-in-use (*VU*). But he opposes vigorously Smith's definition of exchange value, according to which "the things which have the greatest value in use have frequently little or no value in exchange and on the contrary,"[29] because in this definition a value is clearly dependent on the demand and supply structure. Smith's conception of values amounts to founding values and utilities on the changing preferences and demands of consumers; consequently it is the market which may 'create' or increase and decrease values independently without any equivalent increase or decrease of labor. But this leads unavoidably to alienation. The reverse, a microeconomic theory as outlined by Marx in *Capital*, based on the avoidance of alienation, implies, as we shall show, a completely different foundation of microeconomics including an ethical 're-education' of all participants. The following discussion is perhaps the only existing argument for the Kuhnian thesis that paradigms influence the development of scientific theories, for we have shown how Marx's microeconomics and social philosophy is predetermined and thoroughly influenced by the paradigm of alienation. For Marx, under the influence of the alienation paradigm, labor values were always more fundamental than prices. He thought that values could be defined entirely by labor and, in a modern sense, by technology, and that these values should never be influenced at all by changes in the wages and prices of the market.[30] Thus Marx's definition of values is clearly conceived under the influence of the alienation paradigm, but what he is really searching for is an ethical foundation of labor and, indirectly, of economy. We do not share the opinion that Marx's labor theory of value is solely a social or an economic problem: it is primarily an ethical problem. This is very clearly expressed when he defines a value in his main definition ethically: "all that these things now tell us is that human labor power has been expanded in their production, that human labor is embodied in them. When looked at as crystals of this social substance, common to them all, they are values".[31] But Marx also provides a slightly different definition, one more socially oriented. "We see then that that which determines the magnitude of the value of any article is the amount of labor socially necessary, or the labor time socially necessary for its production."[32] In the first, more ethical value-laden definition, the laborer adds a fresh new increment of value to the old value by additional labor, no matter what the specific character and value of that labor may be. Thus we get

7.1.4 $\Delta V = \Delta l = x_1 w_1 + x_2 w_2 + l_{12}$,

where x_1 is the amount of the raw material x_1 and x_2 the amount of the raw material 2; w_1, w_2 the corresponding values and l_{12} the value of the hours of labor needed. Since for the capitalist the value of the end-product is always greater than the sum of the values of the commodities x_1, x_2 used in the production, the capitalist acquires this surplus value from l_1 according to Marx.[33] If we analyze Marx's second definition, we get a more socially defined value. According to Leontieff's input-output analysis, one unit of the net output of x_1 is obtained by producing q_1 units of gross output of x_1 and q_2 units of gross output of x_2, so that l_1 q_1 hours of labor are employed in the x_1 industry and $l_2 q_2$ in the x_2 industry. The labor time socially necessary for production of one unit of s_1 is, therefore, $\Delta l'$ and its value $\Delta V'$.

7.1.5 $\Delta V' = \Delta l' = l_1 q_1 + l_1 q_2$

The question is, is $\Delta V = \Delta V'$ and $\Delta l = \Delta l'$ or not? Is Marx's ethical value definition equivalent to the social one? According to Morishima this question can be answered positively, if we assume that our economy is a simple commodity production economy. Thus we are clearly thrown back to the principal question: does there exist an interpersonal comparison of labor values with value-in-use (exchange values) and in-between labor values on the basis of an ethically justifiable equivalence between labor times of different workers doing different kinds of work? The positive answer for the first question is important, for it reintroduces Marx's labor theory of value into modern micro- and macroeconomics as a first step to solve the ethical or the alienation problem. Therefore, we do not share the opinion of most of the economists and philosophers, e.g., Popper's, who rejected Marx's labor theory of value as economically unusable. But we think that its ethical significance is far more decisive for Marx himself and our society, especially in a 'small is beautiful' economy (Schumacher). Morishima has adduced conditions (which we have slightly modified) which re-establish economically Marx's labor theory of avlue. These conditions are:

7.1.6. In a simple economy, producing all kinds of goods, e.g., capital goods (means of production) and wages or consumers' luxury goods, the following (ethical) restrictions should hold.

7.1.6.1 First, in each industry one and only one method, namely the best method, of production should be used.

7.1.6.2 Second, each industry has to produce one kind of output without any byproducts, e.g., pollution.

7.1.6.3 Third, there are no primary factors of production other than labor, and labor time, which is, on the one hand, measured by an objective ethical standard and, on the other hand, consists of contractually fixed interpersonal units of labor time. This is the ethical equivalence condition.

7.1.6.4 Fourth, all capital goods should have the same span of life (unity period of use).

7.1.6.5 Fifth, all commodities should be produced in the same period of production (convention about unit time).

7.1.6.6 Sixth, each production process is of 'input-output' type; inputs are made thoroughly at the beginning and outputs at the end of a period, and labor is used only once in each production period.[34] Morishima has shown that a revised labor theory of values enables us to construct a von Neumann-like macroeconomic model which is a proof that Marx's labor theory of value is economically feasible and a practical one. But the conditions required to put this into practice would change our present micro- and macroeconomic system. It would replace a competitive economy by a moderately planned economy, and would use the debatable ethical equivalence condition 7.1.6.3 which we will discuss later. It would avoid alienation and introduce an ethical 'distributive' standard unit of payment for wages, provided we can find a solution for condition 7.1.6.3 which, in an affluent, rich society, may be higher than in an underdeveloped poor society. This highly utopian economic system would be based on the radical ethical equivalence condition 7.1.6.3. According to this condition not only are all human beings equal, but their labor values, i.e. wages for labor, measured in interpersonal temporal standard units, are strictly equal for the same periods of labor time. Briefly: All should receive the same wages for the same labor time.

8. MARXIAN PARADIGMS VS. MICROECONOMIC STRUCTURES

After having discussed the agnate sets, i.e. the background knowledge of the Marxian and the standard (capitalistic) version of microeconomics, we want to compare the full interpretation of $\langle K, I^C \rangle$ and $\langle K, I^M \rangle$. We shall use two sets of semantical interpretation rules which are influenced by the two agnate sets, consisting of the paradigms and, of course, of the corresponding maxims, principles, and of the superhypotheses of the background knowledge.

8.1 *Interpretations of Axiom 1*

In the standard interpretation, the set N is interpreted as participants in the market: consumers, producers, laborers, or owners of means of production. The standard paradigm 4.4 will permit a dynamic and changing formation of classes in cooperative games or in non-cooperative, competitive games, e.g. formation of unions, of oligopoly, duopoly or monopoly. But the Marxian paradigm of alienation 5.4 entails only a strict competitive two-class society of workers and capitalists (producers). Whereas in the capitalistic standard interpretation all participants of the market will maximize their utilities (profits) according to maxim 4.2, in a Marxian, albeit highly unrealistic, interpretation, the worker has to go through a social, historical or demagogical learning process before he may adopt step-by-step maximization of his individual utility in the form of higher wages and even has to learn group behavior and adopt group rationality. This group 'learning' occurs only after he is cognizant of alienation, i.e of the way in which he is being exploited by the capitalists. The laborer will do this, according to Marx, in order to escape growing alienation. We see clearly that at the moment the workers suffer from the influence of alienation, they begin to interpret the structure K as exploitation and will start maximizing their utilities. But paradoxically, if we would abolish alienation and exploitation of workers, the workers would be freed from capitalistic influence and have to stop maximizing their utility (e.g. income). Therefore, it is the alienation paradigm which forces the workers and Marx to become social revolutionaries and propagators of the struggle of classes. The capitalistic standard interpretation, on the contrary, is not exclusively based on class struggle. Rather, it is game- or decision-theoretically oriented and consequently we are more or less indifferent to class struggle if we use this interpretation. This interpretation offers, instead, a continuum of social and economic solutions of the cooperative, the discriminatory, and the non-cooperative kind, in which the merciless class struggle is only one extreme form. Theoretically speaking, only the solution of the extreme form of class struggle would be that of a zero-sum game.

8.2 *Interpretation of Axiom 2*

In our capitalistic, standard interpretation *X* are raw materials, commodities, goods and end-products insofar as they belong to production or exchange of goods. In a similar way Marx himself regarded firstly the economy as "an immense accumulation of commodities"[35] in their physical or natural form. But this is not really important for him.[36]

8.3 *Interpreation of Axiom 3*

Marx is more concerned with the value form of commodities, since for him economics is the science and practice of values. But the value of a commodity, as already mentioned, is nothing other than, according to Marx, the "embodiment of one identical social substance, viz., human labor",[37] or human life. Thus the interpretation of structure *K* by means of the alienation paradigm and the labor theory of values begins very early. In the capitalistic standard interpretation, values (*W*) are utilities which are obtained by the functions *f, c*, which are again expressions of the characteristic supply and demand form of the market combined with uncertainty and risk in decision-making. But in Marx's interpretation, values can be created, changed, and modified only by human labor. This interpretation is faithful to the paradigm of alienation. More and more, we notice that the two interpretations amount to different foundations of value theory and two different foundations of economics. Since Malthus, Smith and Ricardo, the standard interpretation of *W* has been based on value-in-use, *VU*, the forerunner of our present utility theory of econometrics, based on preference. In our case, we have proposed this by superhypothesis 4.1 (an interval utility theory which is, in itself, an axiomatic structure described and axiomatized elsewhere by the author).[38] Thus prices, *P*, costs, *C*, profits, *W*, are monetary values obtained by specific, continuous utility functions *f, c, p* (which are again, as already mentioned, dependent on the underlying dynamic-preference structure of the market, as well as on the costs, wages, according to axioms 4–6. Marx, on the contrary, regards the supply and demand substructure as a constant value-in-use, which, in fact, can be neglected only in the case of a market equilibrium. He assumes vaguely that always some kind of equilibrium of exchange exists on the market. Because of the dependency of the value of a commodity on the price of the market (expressed in axiom 5), we may speak of the price of the commodity, according to our capitalistic standard interpretation, generally as the value-in-use (*VU*). But Marx has to reject vigorously this determination of a value: "the mystical character of commodities does not originate, therefore, in the use value".[39] Marx counters dialectically the capitalistic value-in-use conception with his labor theory of value as an 'objective' value theory versus the 'subjective' one (as pointed out recently by Becker).[40] But the subjective theory of values in its present form is, according to the Western standard interpretation, actually based on individual preferences and demands, as demonstrated in *K* and is the cornerstone of modern microeconomics. We may therefore ask: Was Marx able to found his labor theory of value really independently of value-in-use and modern utility theory based on preferences? Firstly, it is well known that Marx did not differentiate at all between prices and exchange values, therefore we may assume that both are equal in Marx's *Capital*. Secondly, Marx's argumentation is strictly dialectical: as reported by all Marx's interpreters, value-in-use is the thesis, value or labor value the antithesis, and the synthesis is alternatively either the exchange value or equivalence of values or value comparison of abstract labor values. The following diagram, which should help to support this interpretation of Marx's labor theory of value, deals with the different forms of equivalence of values which Marx has used and which is the general form of a value or value comparison according to him (the following numbers refer to Marx's *Capital* (Engel's edition) (79)). It can be used only in exchange situations or when we compare commodities A and B. Exchange values and comparison exist only when trading is given, expressed by the strict equivalence '=', whereas a difference of values, expressed by '>' can only be caused by labor (59), not by preferences. But what is the real meaning of the equivalence sign? On the left side of Marx's value equivalence, we have values-in-use (*VU*) (56); on the right side, abstract values or labor values (*LV*). Marx uses this very important equivalence for value comparison of all commodities. The value-in-use is called by him the relative value (56) and is variable, while the right side is called the equivalent value and is fixed (56). We find the following comparison:

8.3.1 Equivalent form of values

VU =	*V* or *LV*
value-in-use (56,71)	value (labor value) (56)
relative value (56,71)	equivalent value (56)

abstract form of equivalence:

x commodities A	y commodity B

concrete forms of equivalence:

20 yards of linen	1 coat (74)
1 coat	20 yards of linen
10 lbs. of tea	2 ounces of gold
1/2 ton of iron	money (75,80)
carrier of value = commodity: linen	carrier of value: coat

medium of comparison: actual exchange and value comparison (34)
Material standard unit: Gold, money (80); Ethical standard unit: unit of an interpersonal standardized labor time embodied in the created commodity (thing) (45) which is identical with a quantum of laborer's life-energy.

See: 48, 52, 53, 57 60, 63, 64, 65, 66 71, 76, 81	See: 52, 53, 58, 59, 61, 63, 64, 66, 67, 71 (See notes)

The relative value or the value-in-use (*UV*), and value or labor value (*LV*) "are initially exclusive, antagonistic extremes", but Marx's economic system hinges on the possibility of comparing both. He calls the antagonism of both, if seen only in a material form, the fetishism character of commodities.[42] Therefore we have to regard it value-theoretically. Marx's thesis-antithesis character of value-in-use and labor value will help us to understand the difference between both, and why our whole econometrically founded microeconomic structure K on the one side is economically and ethically incompatible with the labor value theory. There is no dialectic synthesis of both, in spite of all Marxist interpreters. Since we are unable, as we shall demonstrate, to use labor and labor time as the real interpersonal measure and ethical standard of values, we shall see that Marx has done this with the help of the exchange value alone and never with his labor theory of values. Since in the foregoing section, we have already derived the labor theory of value from the alienation paradigm ($PA \rightarrow$ LThV), we face now in the first chapter of *Capital* the dramatic climax in the drama of the confrontation: microeconomic structure versus Marxian paradigms. It is, of course, not astonishing that for this argument Marx has used the vagueness of the dialectical method, which cannot be used in this case for any practical purpose (see also E. Mandel and Becker).[43]

But how did Marx try to overcome these difficulties? In the first step, the value-in-use and the labor theory of value in the form of an abstract 'ethical' value are, in fact, according to Marx, in a relation of a dialectical equivalence only in the state of actual exchange or comparison, never *per se*. This means the basis of the equivalence 8.3.1 is actual exchange (M). Thus, we have x commodities of $A = y$ commodities of B, viz., Marx's famous example of twenty yards of linen = one coat. A is the relative value or the value-in-use, and B is the equivalent value or later the abstract labor value. Both are, according to Marx, antagonistic extremes and 'exclusive'.[44] B, or the body of the commodity B that serves as the carrier of the equivalent value, "figures as the materialization of human labor".[45] In generalized form, according to Marx, the equivalent or fixed value (B) is abstract labor, which always contains the concrete labor, e.g., the time of production of a coat. It finally is converted into conventionally standardized money which serves, as Marx cautiously expresses, only as a value carrier and equivalent for exchange. This expresses finally, according to Marx, a value equivalent (wage) for the labor time of manufacturing twenty yards of linen by a single worker and manufacturing a coat in terms of money (80). Astonishingly enough, the whole problem of an ethical interpersonal value comparison, according to Marx, is now solved in a non-monetary sense, e.g. the value-in-use standing on the left side of the equivalence is expressed by the labor value, standing on the right side. But without any fixation of an interpersonal standard for a unit time of labor and without justification of how we should compare different kinds of labor, the ethical condition which should prevent alienation is unrealistic and will not work. For example, if we would compare a 'vivid' labor value (wage) with a dead labor value, i.e., a laborless value such as capital gained by interest, then the laborer would be alienated by the fact that he is working and the other one is profiting from less work or not working. The worker is alienated, according to Marx, in a double sense; he is alienated from his product (estranged labor), and he experiences an inner psychic alienation vis à vis a person who profits without working or with less work. But one could see this completely differently: the workers are far more alienated if there exists no interpersonal, ethically justifiable comparison of working time for all different kinds of labor: skilled, unskilled, intellectual, manual. This will become a salient point of Marx's ethical foundation of economics. The main difference between the capitalistic standard interpretation of values as utilities and the Marxian one of values as labor values will become more distinct in the next interpretation. But in any case, Marx cannot solve the problem of an ethical standard unit of value for all kinds of labor; instead he falls back into the alienation model of labor which he uses as a threat or a critical and cynical weapon against any nonethical and alienating comparison of values. But he never tells us how to compare the labor values ethically or practically. Comparison of labor time alone does not work at all because it leads inevitably to alienation if we, for example, equate one Einstein-hour with 5 laymen-hours. More of the essential features of the Marxian labor theory of value will be presented in the next interpretations. We will not ventilate here the consequences should an interpersonal, more conventional standard value for time units of all kinds of labor exist, which would mean simply: equal money for equal labor time for all, or the more specialized labor, the more merit. There exists no society where a skilled worker will not earn more than an unskilled one. The unsolvability of interpersonal comparison of the earnings of a genius with the earnings of an average worker will create alienation again and again. Therefore, it seems plausible that Marx had to turn to the utopian revolutionary solution to create a future society without private property and no maximization of utility and greed, where this ethical problem simply no longer exists. But nevertheless, we regard Marx's labor theory of value as an original contribution to an ethical foundation of societal labor, societal production and economy which amounts finally to a re-education of the masses, where they have to learn to forget about maximization of external utility. Until now it has been impossible to give Marx's ethical value equivalence a societal and ethical reality. (Maybe it will be possible in a future society, where all the primitive labor is done by perfect robots and microprocessors permit everybody to work two hours a day at home). Alternatively, in the capitalistic standard interpretation, the interrelations between costs, prices on the market and profits, are of an empirical nature, they can be understood easily and can be described mathematically, i.e., game-theoretically. It seems to be easier to put the behavior of people in the market under 'ethical restraints' or legal rules than to compare their actual work or labor in a just and ethical sense or to strive for a revolutionary, more just redistribution. All this leads to the present attempts to found economics ethically or on welfare economics, but even these attempts will never completely ban alienation from the social and economic scene.

8.4 *Interpretation of Axiom 4*

The standard, as well as the Marxian interpretation, does not differ with respect to the interpretation of the set C as costs. In Marx's own formulation in *Capital*, he terms 'c' the constant capital and he terms the wages the variable capital in his symbolism k.[46] We may note that for the sake of simplicity in our axiomatization, the cost function is equivalent with the supply function.

8.5 *Interpretation of Axiom 5*

The market in the standard interpretation is the prototype of a value- and decision-oriented structure which increases or decreases values (prices or public values – P_i) automatically. Production, selling, buying of goods, and labor are actions or strategies of the participants, by means of which they may influence the dynamic holistic system, the market, for their utility and profit. (We understand a holistic system in the sense of system theory.) In our standard interpretation, the participant is aware that if the price of the commodity increases on the market, the individual consumer will not continue to buy in undiminished quantities. Therefore he will include this property into his strategy. That means the demand curves, defined mathematically by axiom 5, express the fact that a fallen price will increase demand and vice versa. Axioms 4 and 5 define, according to Malthus, who termed it a universal principle, that for each commodity some price must exist that will 'cause' the supply and the demand for any commodity to be exactly equal. Any departure from such an equilibrium price seems, therefore, automatically corrected by market behavior, by the market pushing the price towards equilibrium. Despite the fact that equilibria are manipulable by speculation, etc., we may regard the microeconomic structure K partly as self-regulating machinery, which nevertheless can be partially influenced by the decisions of the participants on the market. This interpretation is clearly influenced by our Western paradigm of microeconomics, which regards the market as a game with open strategies, moves in the economic game whose rules define the admissible moves and strategies of the players. We assume that each of them is solely interested in winning according to our maxim 4.2 and avoiding losses in the final outcome. It is of the highest interest that Marx, under the influence of the alienation paradigm and the labor theory of value, would have flatly rejected this game-theoretical supply and demand interpretation. In spite of the fact that his forerunners Malthus, Ricardo, and Smith used this interpretation, Marx follows consistently only his alienation paradigm.

Marx regards the supply and demand interpretation of axioms 4 and 5 as a confusion of value-in-use which is not determined by labor with the exchange value determined by labor.[47] According to Marx, what we do in the market is only to equate different kinds of human labor. Here again Marx confuses the ethicist with the economist.[48] For Marx, only the material variety of commodities is an incentive for exchange.[49] Material variety boils down to the fact that some goods are scarce, some not. This 'scarcity' has been used by the Austrian Neomarxist school to save the Marxian economic system. According to Marx, exchange on the market cannot create or change values,[50] neither can exchange give an incentive to buy cheaper or sell more expensive commodities.[51] Creation without labor would be creation out of nothing, and buying would be "an act of production"[52] of laborless creation of profit or surplus value. If this were so, a capitalist could create values out of nothing and violate Marx's ethical value equivalence. The market would be a value-creating *perpetuum mobile*. But Marx rejects this idea. In the same sense as in the second principle of thermodynamics, the creation of energy out of nothing is rejected, so the alienation paradigm and Marx's labor theory forbid the laborless creation of values. The reason again is that it would finally lead the worker, i.e., the person who creates, into a social alienation facing another person using laborless values. Marx's rejection of axioms 4 and 5 is clearly motivated by a socio-ethical prophylaxis. More and more, the alienation paradigm is put against the microeconomic structure and is, from now on, the unavoidable collision course, which begins with Marx's troublesome equation of value-in-use with labor value $VU = VL$ in the moment of exchange or when they are combined in a 'dialectical synthesis' of being exchanged. It is clear that Marx presupposes here three functions which we shall introduce here. One we shall call the labor function l, which allots a continuous value (increment) to the already existing value-in-use of raw materials, the other one, the exchange value function which we symbolize by m, and the use value function, by u. To support the collision course, we want briefly to compare in a formal way the individual presuppositions of l and m and use values in order structures of Marx's use value and exchange value, and exchange value and labor value. From a modern point of view one may regard general utility or value theory as a coupling of different order structures. The finer and better the determination of the value-differences of the values, the finer and more complicated the order structures of the utility or value theories involved. From this point of view, value-scales are nothing other than axiomatic structures which define the degree of fineness of the order of the values (utilities) and guarantee, at the same time, the isomorphic or homomorphic mapping of preferences on values (utilities) by means of a representation theorem. One may wonder, therefore, which kind of structure and order the Marxian value-in-use theory, as well as the Marxian labor theory of values, might have.

We will base the following considerations on three definitions, which express value-in-use, exchange value and labor value in a more familiar notation:

D1: '$\succcurlyeq$' defines the weak preference, i.e., 'at least as good as'

D2: '$\succ$' defines the strict preference: $x_i \succ x_j =_{df} x_i \succcurlyeq x_j \;\&\; -(x_j \succcurlyeq x_i)$

D3: '$\sim$' defines the indifference: $x_i \sim x_j =_{df} x_i \succcurlyeq x_j \;\&\; x_j \succcurlyeq x_i$

We introduce here, for establishing comparison for all cases, '$\succcurlyeq$' for 'has at least as great use value as', or 'has at least as great exchange value as', or 'has at least as great labor value as', i.e., a general concept of comparison. Thus we are able to compare the different values. Marx's value-in-use concept imposes, therefore, on the commodities x_i, x_j, $x_k \in X^U$ where $X^U \in X$ is a certain general order which we will define in the sense of Smith, Ricardo and Marx by the next definition.

8.5.1 D.4: $VU = \langle X^U; \succcurlyeq, \sim \rangle$ is a structure of the value-in-use if and only if the following conditions are fulfilled:

8.5.2 $\succcurlyeq$ is a quasi-ordering relation, i.e., reflexive and transitive;

8.5.3 X^* is the set of equivalence sets in X;

8.5.4 There exists a one-one mapping function u, whose domain is the objects X and X^* and whose values are values-in-use (utilities);

If these conditions are fulfilled, then the subset X^U of X is 'value-in-use ordered'. We see immediately that our axiom 8.5.1 introduces a quasi-order in terms of strict preferences just in the same way as it is done as a precondition in any ordinal or interval and higher utility theory (see Leinfellner, axiom 1).[38] Therefore, Marx's value-in-use concept is actually an ordinal type of scale. From the point of view of order, the order of it is less than that of an 'interval type' of utility scale (*SC*). Therefore, any interval type of value scale entails value-in-use theory, i.e., $SC \to VU$, but not vice versa. We define now the structure of the labor theory of value (LThV) in the sense of Marx.

8.6 *Labor Theory of Value*

LThV $=_{df} \langle X; \succcurlyeq, \sim, l \rangle$ is a structure of a Marxian labor theory of value if and only if the following conditions are fulfilled:

8.6.1 $\succcurlyeq$ is reflexive and transitive;

8.6.2 The above standing definition 8.5.3 holds;

8.6.3 X^* is a set of equivalence sets in X;

8.6.4 l is a real-valued, continuously mapping function whose domain is the time units used in the productions of the goods x by a certain laborer i and whose values are real-valued numbers or abstract labor values;

8.6.5 m is a real-valued, continuously mapping function, whose domain is the goods or commodities on the market and whose values are real-valued numbers or money;

8.6.6 Ethical equivalence condition: $i = j$ or the labor values for worker i and j are interpersonally exchangeable, i.e., $l_i(x_n) = l_j(x_m)$ if $t_i x_n = t_j x_m$.

If conditions 8.6.1–8.6.6 hold, then the following relations exist between Marx's labor theory of value, Marx's exchange values and Marx's value-in-use, for $t_i = t_j$, $n = m$ or $n \neq m$

$$LV \leftrightarrows EV \to VU$$

or

$$l_i(x_n) = l_j(x_m) \leftrightarrows m(x_n) = m(x_m) \to u(x_n) \sim u(x_m);$$
$$l_i(x_n) > l_j(x_m) \leftrightarrows m(x_n) > m(x_m) \to u(x_n) \succcurlyeq u(x_m).$$

Given the conditions 8.6.1–8.6.6 in an ideal society, Marx would have achieved his goal, an equivalence of labor value and exchange values which have both a strict ordering defined by transitivity, asymmetry and completeness. This would impose on Marx's labor function similar conditions as on our money or exchange function. Since both are stronger with respect to order than the use value we have only one arrow in one direction from *EV* to *VU*; $EV \to VU$, but not vice versa, which actually expresses Marx's view that they are opposites. But they are not independent (we have already discussed why). Condition 8.6.6 is highly utopian. If we regard the exchange values as prices, then we have to face the reality that in the markets, wages – the equivalent to labor values – are commodities like any other goods and are regulated by demand and supply. We know that, according to Marx, this can be the source of alienation. It is clear that the labor theory of value and our prices as defined by axiom 5 lead to the same order, if we replace the labor function l by our price function f. Therefore, they may be regarded as order equivalent, but they are of a completely different nature.

It is clear from the above that: 1. Throughout *Capital*, Marx has confused exchange value with prices, because they have similar order structure; 2. Both are completely different with respect to their origin; 3. In addition he is wrong in regarding exchange values and use values as independent, though they are certainly of a different order. All this supports 4. our assumption that Marx's assertion that there is a 'dialectical' synthesis between value-in-use and labor value is wrong. It is clear from 8.6.4 that Marx's definition of a commodity having exchange value is just another formulation of being exchangeable by means of money, which leads by means of assumptions 8.6.4–8.6.6 to the peculiar definition of prices in *Capital* by means of labor (l) alone. But, contrary to Marx, our functions f and c, defined in axioms 4 and 5, are far more complicated, and are far more socially dependent. They are market-dependent and are, in fact, a function of the supply and demand behavior of all participants on the market. It is clear that Marx has no other choice but to found exchange value on labor value. This explains the abrupt introduction of the labor theory of value into *Capital*, done solely under the influence of the alienation paradigm. This also explains Marx's tendency towards a political economy, where social relations such as labor are the fundamentals of society, but never the relations between goods, commodities and men as is the case with our general preference relations. Our short study and reconstruction of Marx's labor theory of value and its relation to exchange value and value-in-use has shown very clearly that Marx has imposed his labor theory of value artificially on microeconomics as well as why he himself regarded his labor theory of value and value-in-use as irreconcilable extremes. If we begin with our hypothetical labor function l in Marx's labor theory, then we may derive theoretically the exchange value and value-in-use, but we cannot get practically from value-in-use to exchange value or labor value, i.e.: $LV \nleftarrow EV \nleftarrow VU$. Maybe that was the reason why he regarded both as independent. Marx himself was fully aware of this because he defined value-in-use sometimes in the sense of preferences, sometimes not: "All commodities are non-use values or exchange values for their owners and use values for their non-owners".

Marx has imposed his labor value theory on economics solely for the purpose of founding it ethically and to avoid alienation. But by doing this he could never explain how preferences influence exchange values in the sense of $VU \to EV$. But exactly this has been achieved by modern utility theory under risk and uncertainty.

Instead of using the preference interpretations, Marx expresses the prices in the market as exchange values of commodities in terms of something common to them all, i.e. labor (see 8.6.4) because he regards labor as a socially more important relation than exchange or preference. Furthermore, only an ethical justification of labor in the sense of the alienation paradigm guarantees him an ethical foundation of society as well as of economy and the avoidance of alienation. But l and f are different, since the price function f and Marx's labor function l differ fundamentally as has been shown in axioms 5 and 6 and in section 7. Therefore, Marx's labor value theory is superimposed artificially on the microeconomic price structure for the sake of an ethical foundation, but seems to be incompatible with it.

8.6.7 Interpretation of axiom 6. The interpretation of W is that of profit in any standard interpretation. Profits are values, utilities obtained by the difference of the prices on the market (the public values) minus the costs (the labor values) of the specific commodities. This represents a strategy, because the costs of production can be influenced by the producer, according to Marx, by decreasing wages. Thus, maximization of profits can cheaply be obtained by reducing wages, i.e. by exploitation of the workers or unemployment, if and only if they cannot neutralize the strategies of the producers by counter-strategies, e.g. by laws, legal measures and strikes. The same holds for price manipulations, since producers and workers are basically free in their actions (according to the standard paradigm, they may counteract against manipulation of prices according to the rationale of a game). Basically we see that the values such as prices, profits, and the values of commodities can be changed, increased, decreased and even created in and by the market or by interaction between the market and the actions of the participants in the market, even without any labor. But our microeconomic structure is not immune to alienation, e.g. people feel alienated by rises of prices without increasing costs. The question is, should we follow Marx and because of that abolish our microeconomic system $\langle K, I^C \rangle$?

Marx identified profits with surplus value, but consequently it is strictly forbidden for him and for all his followers to explain the profit and the surplus value by means of the supply and demand substructure of the market. We quote: "the creation of the surplus value and therefore the conversion of money into capital can consequently be explained neither on the assumption that commodities are sold above their value, nor that they are sold below their value",[53] i.e. surplus value can only stem from alienated labor. Therefore, we have only one choice, that of alienated labor, i.e. that microeconomics comes inevitably under the dictate of the paradigm of alienation. That is, according to the labor theory of value, surplus value, or profit, has been created only by labor (see 7.1.1–7.1.3). Profit, therefore, is labor, but from his ethical point of view is called "dead labor".[54] Marx even would have agreed with our formulation of axiom 6 if we substitute for $W(x_i, x_j)$ in axiom 6 Marx's own symbols m (which is a surplus value),[55] and replace $x_i f(x_i + x_j)$ by Marx's own symbol w (exchange value), and $c(x_i)$ by k (Marx's symbol for costs). Doing so, we find in *Capital* the same formal expression: $m = w\text{-}k$, but with a different meaning. This is proof that Marx actually acknowledged the same microeconomic structure K which we axiomatized here. The difference results only from the influence of the alienation paradigm and Marx's labor theory of value with respect to the interpretation of axioms 3, 5, and 6. Therefore, Marx ends up with the paradoxical formulation: "the profit of the capitalist is due to the fact that he offers something for sale for which he has not paid anything". Marx finally states the famous formula: surplus value/value of labor power = unpaid labor/paid labor.[56]

8.7 *Interpretation of the Theorems*

8.7.1 T1, if interpreted, depicts and explains early capitalism, described extensively by Marx and Engels. This early capitalism was a Robinson Crusoe economy, where in fact the producers, i, j, maximize their profits without regarding anything else. Capital accumulation and exploitation of labor sets in. Our theorem 1 offers us a solution of the basic economic market situation in the form of a behavioristic rule. From the point of view of our theory, if the agnate set B^C is assumed, a strategy for the duopolist 1 or 2 is prescribed. Therefore, we may use a deontic operator 'O' which means 'you are obliged with respect to the agnate set B^C to do . . . . ' before the empirically interpreted theorem. Thus, the theorems adopt the character of advice, prescription or instruction. Although each of the producers knows that he is dependent upon the market behavior (for example, increase of production) of other producers, he may game-theoretically assume that the other producer will not change his strategy if he changes, for example, his own output of production. If each of the producers makes the assumption that his strategy will not influence the others, e.g. x_j is constant in the eyes of the duopolist i, vice versa, then each of them should behave as maximizer of his profit, that is, i maximizes his profit:

$$\mathrm{d}W_i/\mathrm{d}x_i = 0$$

and j:

$$\mathrm{d}W_j/\mathrm{d}x_j = 0,$$

and each acts as if he alone were in the market.

8.7.2 T2 explains the leading role of equilibrium in Marx's economic theory, especially of Pareto equilibrium (EQ), where each producer's profit is dependent on the other producer. This theorem was used by Marx to explain the transition phase of capitalistic development into industrial societies. Theorem 2 expresses an equilibrium of profits and production costs. In such a case, each change of profit is dependent upon a change in production costs. We call this kind of equilibrium Pareto's (closed) equilibrium. Certainly we cannot assume that each duopolist, or oligopolist, has this partial differential equation present in his mind, but it is certainly a mathematical representation of his performance. We regard theorem 2 as a mere theoretical expression, explaining the growing interdependency of 'capitalists' as described by Marx.

8.7.3 T3 explains the possibility of monopolization and trust formations as a consequence of K. Since each duopolist and oligopolist has his own free will, we have to take this fact into consideration. Our microeconomic structure K provides an answer, simply by introducing two cases of free decisions. The first case is the case of cooperation. Both producers decide to cooperate and to maximize the common profit. They form a tacit understanding or public trust (or cartel). This trust tells us how both will divide the common profit in the best way:

$$\max \{ W^*_1(x_i, x_j) + W^*_j(x_i, x_j) \}$$

But if either one of them decides not to cooperate or to negotiate, but chooses free competition, both will fight for the market. This situation is described theoretically by the next theorem.

8.7.4 Interpretation of theorem 4 (the innate temptation to gamble). In this interpretation, the powerful influence of the 'game' paradigm becomes obvious. The offered solution is a typical game-theoretical one. If both (all) firms refuse any negotiations and decide for the free competition, the first firm will win what the second loses, or the second firm will win what the first has lost. This is a characteristic feature of a two-person zero-sum game. If the first producer tries to enlarge his profit, he has to maximize the difference $(W_i - W_j)$. In exactly the same manner, his adversary will react by trying to make the difference $(W_i - W_j)$ as small as possible. This is the only way to react if both have decided to fight economically. Especially if the cost functions are different, a little firm which produces very cheaply can win the competition against a big firm with high costs of production. There is, according to theorem 4, a stable solution to this competition. Even if both are not willing to cooperate or to begin bargaining, it is best for both to act in the following manner: If the second firm will minimize the differences, the first firm can take the maximal values of the minimal differences. Then there is no choice left for the second, he must take the minimal value of the maximal differences. In other words, if both firms have only two strategies, to raise or to limit the production, and if the available differences are given by the monetary values \$4, \$3, \$6, \$5 per 100 units, being the profits of a commodity x for the first firm and, therefore, \$4, \$3, \$6, \$5 losses for the second, we have a situation which may be theoretically represented by a matrix. If the first firm uses the first strategy and limits its production, and the second firm uses its first strategy as well, the first firm will win \$4 per hundred units of a commodity x and the second firm will lose \$4. In case the first firm uses its first strategy and the second firm uses its second strategy, the first firm will win \$3 and the second firm will lose \$3. (The second strategy is for both firms to limit production.) If now the first firm limits its production, chooses its second strategy, and the second firm uses its first strategy, the first firm will win \$6 and the second will lose \$6. In the case that the first firm chooses its second strategy, and the second firm its second strategy too, the winnings for the first firm are \$5 and for the second firm the losses are also \$5. We form now the above-described matrix, which is equivalent to the rationalization of the game. We do not demand that the participant is rational, but only that he has an insight into the game. Game theory offers such an insight. After understanding the game and the rules, the gambler will take his opportunity and maximize his utility.

The so-called "minimax theorem" of game theory can be applied within

		Firm 2	
		1	2
Firm 1	1	\$4	\$3
	2	\$6	\$5

our theory of the market; it will explain competition and declining rate of profit, as Marx has described it.

In this $\langle K, I^C \rangle$ theory, the general principle of Pareto optimality, or equilibrium, is introduced stepwise by axioms, which define the necessary functions and lay down, in a formal (mathematical) sense, properties of the functions f, c which enable us to form the first derivative and the second derivative of the functions, especially of the function W_j, the main presuppositions of classical economic theories. Since the axioms guarantee differentiability, there is no need to introduce equilibrium between the profits, the produced goods, and costs separately. We derive P (Pareto optimality) for closed states by theorem 2, the most fundamental construct of our theory, deductively:

$$\mathrm{d}W_i/\mathrm{d}x_i \mathrm{d}x_i + \mathrm{d}W_i/\mathrm{d}W_j \mathrm{d}x_j = 0 = \mathrm{d}W_i \text{ and}$$
$$\mathrm{d}W_j/\mathrm{d}x_i \mathrm{d}x_i + \mathrm{d}W_j/\mathrm{d}x_j \mathrm{d}x_j = 0 = \mathrm{d}W_j$$

by application of the rules of differential calculus. Since we have assumed that the underlying logic is that of set theory, we may include differential calculus in set theory, as is usually done. The most astonishing result is that business men are completely free to react in the market as if each of them were alone (Robinson Crusoe economy) according to theorem 1, or according to the third theorem, they may cooperate and divide the common profit, or compete with each other according to theorem 4. Thus, theorem 1 may characterize the state of an early capitalism where closure of the economic market finally leads to a state of the economy described in a neutral way by theorem 3 (as monopoly capitalism, according to R. Luxemburg) and to a state of economic warfare amongst capitalist concerns or nations as described by Marx. It is interesting that Marx's theory of capitalistic development has an evolutionary dialectical, dynamic super-structure. The state of economy expressed by theorem 1 is the thesis (initial state), and the state of capitalistic economy expressed by theorem 4 is the antithesis (competitive struggle), which is followed by the synthesis, a state of economy (capitalistic monopoly) expressed by theorems 2 and 3.

There are many more features of our market which can be derived formally from our structure K plus the agnate set B^C. The most important one is certainly Marx's famous class struggle. We may derive the class struggle as a theorem from our microeconomic structure K and the standard or capitalistic agnate set B^C by interpreting the characteristic value function ν for groups or classes in the usual game-theoretical way. If $\nu(i) = \emptyset$, which means the actual state of a society that only classes may achieve something, never individuals alone and for the two classes C_C, C_P (where C_C are the capitalists, C_P are the Marxian proletarians), the following holds: if $C_C \cap C_P = \emptyset$, and $\nu(C_C) + \nu(C_P) = \nu(N)$, where N is Marx's two-class society, then this expresses Marx's famous law of class struggle, if C_P are the proletarians or workers and $C_C = N - C_P$ symbolizes the capitalistic class. Class struggle is a strict, competitive game, where the one class loses what the other wins. The strict, competitive fight of the two classes is, in Marx's economic theory of *Capital*, the only and final outcome. There is practically no alternative to this outcome, whereas in the capitalistic standard interpretation of the structure K, there is a variety of other alternatives of competitive bargaining. We may derive the following consequences from our microeconomic structure: in the usual game-theoretical way, the group or collective value of any coalition C, or of cooperative bargaining between the two classes C_P and C_C, is given by the expression: $\nu(C) = \delta \min(|C \cap C_P|, |C \cap C_C|)$, where δ is the difference of the values of the wages offered by the laborer or the labor's union and the value of the wages which the capitalists are willing to pay. The result is clearly that the employers will always hire the employees who are willing to work for the smallest wages. This explains very well Marx's three laws concerning the surplus value.[57] We want to conclude the discussion of the interpretation of the structure K and its theorems. Thus, we have proven that Marx could have taken all his knowledge of microeconomics from this underlying invariant structure, but saw or interpreted it in a different sense under the influence of the alienation paradigm. The final question is: Why did he do this? Was it really ethically motivated?

9. THE MARXIAN DILEMMA AND WELFARE ECONOMICS

In this paper, we come to the conclusion that the alienation paradigm played a twofold role in Marx's social philosophy and economics. The first, 're-volutionary' role was a mere negative one. It permitted Marx to criticize alienated societies because of their ethical abuse of labor whenever he wished to do so, and led finally to the revolutionary solution. The other role, 'the social solution', forced Marx to found economics and his theory of society upon social or ethical principles of labor-value equivalence in order to avoid alienation. Now we regard welfare economics as an old branch of economics which wishes to improve human welfare and social conditions. Therefore, Marx's positive annulment of alienation has to be considered as an ethical approach to welfare economics, among many other possible ones. Interestingly enough, this is the theme of a series of recent works which offer a different solution for an ethical foundation of economics. In this final chapter, we want to compare both solutions of welfare economics. Both solutions agree that microeconomics is, by its nature, more an 'illfare' rather than a welfare economics and that, therefore, decisive steps towards an ethical foundation of economics have to be taken; but these solutions fundamentally disagree on how this result is to be achieved. Marx's choice of the revolutionary solution can be seen as a consequence of his foundering in putting up a socially workable, ethical value equivalence between the labor value and the value-in-use or the public values. The question is, can we ever establish an ethically sound standard unit for labor values, in view of the paramount inequality between the quality of human labor, on the one hand, and the equality of life, on the other? It is a utilitarian truism that specialized labor is more valuable for the individual and for society than unskilled labor, but with respect to the Christian value of the worker's life fractions which skilled and unskilled workers have put in equal labor time, they are equal even according to non-Marxian views. But to pay all workers an equal amount of money for equal labor time would take away all incentives for acquiring special skills and would kill our innate greed to possess more goods, property, etc., the hidden motor of market economy. Maybe this will happen once in a future utopian society, where no technical improvements are any longer necessary, and where the scientific and technological development has come to its final end. But we do not know yet today what kind of compensation or merit for outstanding, above average labor should replace the maximization of material utility. To take this away, together with profits, would simply mean to switch off the instinctive motor of the microeconomic dynamic structure K. It seems that more and more our maximization maxim of utility is the evil, or becomes the original sin of our economic society. But we do not share such a position. We regard the shift towards the egotistic side of our societal life as bad, but we do not view the 'religious liberation' of man which initiated the belief in a free market in a free society as being solely responsible for it.

Microeconomics is still far more 'movement of private property' than Marx assumed. Thus, we get a typical Marxian dilemma: If we have inequality of labor values for work done in the same time, then sooner or later alienation will set in. But if we have an ideal, ethical equivalence between labor values and values-in-use, i.e. utilities, then we take away the motor and its energy, i.e. we may throw overboard our whole present microeconomic system. Marx clearly must have seen this dilemma, i.e. the final incompatibility of the microeconomic structure *K*, plus its maxims and paradigms, and his ethical alienation paradigm. As a result, he opted for the revolutionary adventure of changing society, hoping that a future society without the 'maximization motor' of greed would work. An alternative other than the Marxian solution would be to attempt a radical, ethical and cultural re-education of man. One would need to teach man that he has to give up the maximization maxim, and furthermore that material inequality and alienation are the bitter and brutal consequences of a historical economic system which cannot be changed without giving up the fundamentals of our present social and economic life. But coping with alienation can only be done by: (1) diminishing alienation as far as we can do it; (2) understanding the message of the earlier forms of the paradigm, where the way back, disalienation, is always understood as an inner, individual one, via understanding and insight, and thus reconciling ourselves with the necessary, but brutal, nature of microeconomic life, a hereditary relic of our animal survival strategy of the past. We have to cope with the fact that there will always be material inequality and differences between the members of any society, even if we have reduced alienation to a minimum.

Or maybe the French utopians were right, and our definition of private property has to be changed radically. If property is only of a private nature, if its possession does not inflict harm on others or does not entail any power over the other members of the soceity which is not permitted by the law, then accumulation of private property is in fact no source of social inequality. It seems that the whole system of religion, culture, art, and education with its equality tendency could be used for an inner compensation, in the case where material monetary equality cannot be achieved, provided that alienation and frustration is within socially tolerable limits. Thus, it is obvious that there is a common denominator and trend in the Marxian economic system and in the standard microeconomics of today, which we have worked out above. It is obvious that the common denominator is the trend towards welfare economics or the trend towards an ethical or social foundation of economics or of our microeconomic structure *K*, in order to make *K* better and socially tenable, e.g. diminish alienation.

We may see this trend as the immediate result of the epitheoretical analysis of the underlying invariantly given microeconomic structure K by means of the two interpretations. The present trend towards welfare economics has gotten powerful world-wide impulses from the MIT group, the Club of Rome, the 'green' parties in Germany and support from the ethical foundations of economics by Arrow, Harsanyi, Sen and Leinfellner, which added a new line of thinking to the older school of welfare economics (Walras, V. Pareto, A. Bergson, J. Hicks). Welfare economics, as the branch of economics concerned with improving human welfare and social conditions can, of course, be achieved in this new way of reducing alienation. While the Marxian solution is finally a revolutionary discontinuous one which remains faithful to the paradigm of alienation and to the dialectic method, the new welfare economics must be an evolutionary one, i.e. it must improve its own program and rules.[58] The newest welfare movement or the ethical foundation of micro- and macroeconomics tries, in fact, and contrary to Marx, to maintain a continuous line. It tries to improve the status quo of economics step by step by imposing fair, social and ethical super-rules (or principles) on the microeconomic game. But this new rational approach is only possible after the value-and decision-theoretical foundation and theorization of microeconomics has been achieved. The tremendous advantage of this trend and new foundation of micro- and macroeconomics lies in the flexibility of the value- and decision- (game)-theoretical character of microeconomics, which includes ethical and nonethical outcomes of games. The economy of society as single, dual, and plural (collective) decision-making may be more easily influenced by fair rules and ethical principles which shift the balance from egotistical to more altruistic (social) solutions of a theory of society if it is necessary. This presupposes the rational insight in the market – what we have done here – as well as a rational attitude of the participants, a public or legal constitution of all rules for the future economic 'game' and finally an education to change the agnate sets, which represents the cultural, hermeneutic heritage of economics. Education for fair play can only be achieved by changing step-by-step maxims, principles and the whole paradigmatic aspect of economic science. Thus, it seems that we do not need to change the invariantly given structure of microeconomics, but we have to change the agnate sets or the cultural background which influences the rules of the game and the decision-oriented framework in the direction of a welfare economics.

However, if we take this direction in economics and the social sciences, we follow the spirit of Marx's ethical solution, but in an evolutionary sense. It means that the character of social sciences and economics has to be changed, too. There is no place left for a merely cognitively oriented economics or social science. The quest for a realizing economics, looking for practical, more ethical 'alienation reducing' realizations of theoretical models has to be continued.

University of Nebraska

ACKNOWLEDGEMENTS

For helpful comments on the occasion of giving this paper at the Boston Colloquium, 1975, I thank the commentator H. Gintis, University of Massachusetts and Professors R. S. Cohen, M. W. Wartofsky, and O. Kyn of Boston University. This paper has been supported by a grant of the Research Council of the University of Nebraska.

NOTES

[1] Leinfellner, W., 'A New Epitheoretical Analysis of Social Theories', in Leinfellner, W. and Köhler, E. (eds.), *Developments in the Methodology of Social Science* (D. Reidel, Dordrecht, 1974). Leinfellner, W., 'Epitheoretical Aspects of Statistical Decision Theory', *Proceedings of the Nebraskan Academy of Science* (Lincoln, 1972), pp. 54–55.

[2] Sen, A. K., *Collective Choice and Social Welfare* (Holden Day, San Francisco, 1970). Hansson, B. 'The Independence Conditions in the Theory of Social Choice', *Theory and Decision* 4 (1973), 25–49. Arrow, K. J., *Social Choice and Individual Values*, 2nd ed. (Wiley, New York, 1963). Leinfellner, W. and Gottinger, H. (eds.), *Decision Theory and Social Ethics* (D. Reidel, Dordrecht, 1978).

[3] Leinfellner, W., 'Modelle der Entfremdung: Von der Entfremdungstheorie zur ökonomischen Theorie bei Marx', *Sozialwissenschaftliche Annalen* **2** (1978), p. B1–B23. Leinfellner, W., 'Paradygmaty Marksowskie, Struktur Mikroekonomiczne Metoda Dialektyczna', *Studia Filozofiinauki*, Warsaw 1979, pp. 123–182. Leinfellner, W. and Leinfellner, E., *Ontologie, Systemtheorie und Semantik* (Duncker & Humblot, Berlin, 1978).

[4] Shubik, M., *Strategy and Market Structure: Competition, Oligopoly, and the Theory of Games* (New York, 1959).

[5] Leinfellner, W. and Booth, E., 'Allais versus Morgenstern' in Allais, M. and Hagen, O., *The American and the French School of Economics* (D. Reidel, Dordrecht, 1979), pp. 303–331.

[6] Leinfellner, W., *Einführung in die Erkenntnis- und Wissenschaftstheorie* (Mannheim, B. I. Hochschultaschenbuch 41/41a., 1980), 3nd ed. pp. 145–160. Leinfellner, W., 'Logik und Semantik sozialwissenschaftlicher Theorien', in Kamitz, R. (ed.), *Logik und Wirtschaftswissenschaften* (Duncker & Humblot, Berlin 1979), pp. 163–185.

[7] Ebeling, G., 'Hermeneutik' in Yaling, K. (ed.), *Religion in Geschichte und Gegenwart: Handwörterbuch für Theologie und Religionswissenschaften* **13** (Tübingen, 1959), pp. 242–262, esp. p. 243.

[8] Gadamer, H. G., *Wahrheit und Methode* (Tübingen, 1965), p. xvi.
[9] Sen, A. K., *Collective Choice and Social Welfare* (Holden Day, San Francisco, 1970), pp. 4–46. Schwartz, Thomas, 'On the Possibility of Rational Policy Evaluation', *Theory and Decision* 1, 89–106.
[10] Huizinga, J., *Homo Ludens: A Study of the Play Element in Culture* (Beacon Press, 1955), p. 11.
[11] Weber, M., *The Protestant Ethic and the Spirit of Capitalism* (Scribner, New York, 1958). Troeltsch, E., *The Social Teaching of the Christian Churches* (Macmillan, New York, 1931). Cunningham, W., *Christianity and Economic Science* (1914), Chap. V.
[12] Brentano, F., *Die Anfänge des modernen Kapitalismus* (Vienna, 1916), pp. 117–157. Tawney, R. H., *Religion and the Rise of Capitalism* (Mentor, New York, 1953), pp. 72–115.
[13] Rawls, J., *A Theory of Justice* (Harvard University Press, Cambridge, 1971), pp. 4f. Leinfellner, W., 'Marx and the Utility Approach to the Ethical Foundation of Microeconomics', in Leinfellner, W. and Gottinger, H. (eds.), *Decision Theory and Social Ethics* (D. Reidel, Dordrecht, 1979), pp. 33–59.
[14] James, R., The *Apocryphical New Testament* (Oxford University Press), pp. 411–415.
[15] Grofman, B. and Hyman, G., 'Probability and Logic in Belief Systems', *Theory and Decision* 4 (1973), 179–195.

NOTE: Karl Marx is cited first in an English translation (arabic numbers) and second in the German original version (Roman letters). *The Economic and Philosophical Manuscripts of 1844* will be cited from the edition by Dirk J. Struik (International Publishers, New York, 1967) in the following way: E & PM; the English version of *Capital* Vol. 1 is Engels' edition of the Modern Library edition (New York, Random House, 1906) cited as: 1. *Capital*; Vols. 2 and 3 will be cited from the edition by the C. H. Kerr Company, Chicago, 1925, as: 2 and 3. The German original edition is the *Karl Marx Ausgabe* by Hans Joachim Lieber and Benedikt Kautsky (Cotta Verlag, Stuttgart, 1962). Roman numeral I refers to the first volume of Marx's early writings, IV to the first volume of *Capital* and V to the second and third volume of *Capital.*

[16] Marx, *Capital* 1: (1, 45; IV, 7) Note: the first citation 1, 45 refers to the English, the second citation to the original German version of the first volume of *Capital*.
[17] Marx 1848 (E & PM, 118; I, 573).
[18] Marx 1848 (E & PM, 108; I, 561).
[19] Marx 1848 (E & PM, 108; I, 561).
[20] Marx 1848 (E & PM, 128; I, 585).
[21] Marx 1848 (E & PM, 117, 118; I, 573, 574).
[22] Marx 1848 (E & PM, 111; I, 565).
[23] Marx 1848 (E & PM, 111; I, 565).
[24] Marx 1848 (E & PM, 111–112; I, 565).
[25] Marx 1848 (E & PM, 115–117; I, 569–571).
[26] Marx 1848 (E & PM, 136; I, 595).
[27] Booth, E., 'Foundations of Value Theory and Economics', Dissertation, Lincoln, 1975.
[28] Marx 1848 (E & PM, 136; I, 595).
[29] Smith, A., *The Wealth of Nations* (The Modern Library, New York, 1937), p. 28.

30 Morishima, M., *Marx's Economics* (Cambridge, 1973), p. 10.
31 Marx, *Capital* 1 (1, 45; I, 7).
32 Marx, *Capital* 1 (1, 46; I, 8).
33 Marx, *Capital* 1 (1, 207; IV, 189).
34 Morishima, M., *op. cit.* 3, p. 17.
35 Marx, *Capital* 1 (1, 41; IV, 3).
36 Marx, *Capital* 1 (1, 55; IV, 18).
37 Marx, *Capital* 1 (1, 55; IV, 18).
38 Leinfellner, W., 'Generalization of Classical Decision Theory' in Borch, K., and Mossin (eds.), *Risk and Uncertainty* (Macmillan, London, 1968), pp. 196–210.
39 Marx, *Capital* 1 (1, 82; IV, 47).
40 Becker, W., 'Zur Kritik der Marxschen Wertlehre und ihrer Dialektitk', in Lührs, G., Sarrazin, T., Spreer, F., Tietzel, M. (eds.), *Kritischer Rationalismus und Sozialdemokratie* (Dietz, Berlin, 1975), pp. 201–213, esp. pp. 209–211.
41 Marx, *Capital* 1 (1, 75; IV, 20).
42 Marx, *Capital* 1 (1, 81; IV, 47).
43 Marx, *Capital* 1 (170, 71; IV, 35); Mandel, E., *Marxistische Wirtschaftstheorie* (Frankfurt, 1971), chaps. 1–3; Becker (see 40), p. 206.
44 Marx, *Capital* 1 (1, 65; IV, 19).
45 Marx, *Capital* 1 (1, 67; IV, 32).
46 Marx, *Capital* 1 (1, 235; IV, 220).
47 Marx, *Capital* 1 (1, 177; IV, 154).
48 Marx, *Capital* 1 (1, 187; IV, 50).
49 Marx, *Capital* 1 (1, 178; IV, 156).
50 Marx, *Capital* 1 (1, 182; IV, 160).
51 Marx, *Capital* 1 (a, 183, IV, 160).
52 Marx, *Capital* 1 (1, 178; IV, 156).
53 Marx, *Capital* 1 (1, 179; IV, 157).
54 Marx, *Capital* 1 (1, 257; IV, 245).
55 Marx, *Capital* 3 (3, 49; V, 625).
56 Marx, *Capital* 1 (1, 582; IV, 625).
57 Marx, *Capital* 1 (1, 569–574; IV, 610–616).
58 Leinfellner, W., 'Evolutionary Causality, Theory of Games, and Evolution of Intelligence' forthcoming in F. Wuketits (ed.), *Concepts and Approaches in Evolutionary Epistemology* (D. Reidel, Dordrecht).

HILLEL LEVINE

PARADISE NOT SURRENDERED: JEWISH REACTIONS TO COPERNICUS AND THE GROWTH OF MODERN SCIENCE

I

Copernicus's sixteenth century formulation of a heliocentric cosmos, elaborated during the next one hundred and fifty years through the work of Bruno, Kepler, Galileo and Newton, has been considered a turning point not only in astronomy but in the growth of scientific knowledge and in the history of ideas. The shift from belief in the well-ordered cosmos in which the earth occupies the central position to notions of an expanded universe in which man and his familiar world are relegated to an insignificant corner played a paramount role in the process whereby, as Alexandre Koyré put it, "human or at least European minds underwent a deep revolution which changed the very framework and patterns of our thinking."[1] Even while it was still a subject of debate within astronomic coteries, poets such as Donne and Milton intuited the broader social and religious implications of the altered conceptions of the planetary arrangements.

> [The] new philosophy calls all in doubt . . .
> 'Tis all in pieces, all coherence gone;
> All just supply, and all Relation:
> Prince, Subject, Father, Son, are things forgot.[2]

The declining sense of the rootedness of human values and existing social institutions in the very nature of the cosmos – the process characterized by Max Weber as the "disenchantment of the cosmos" – engendered the anxiety of *Paradise Lost*. Rather than a renewed other-worldliness to compensate for lost paradises, a more worldly orientation begins to develop.[3] Central to that worldly orientation were an increasing interest in nature and a new confidence in the validity of human perceptions of natural phenomena.

Recent studies of Copernicus and reactions to the new cosmology have centered upon two interrelated issues: On the roots of modern science in medieval Christianity and Christendom and, more generally, on the social and cognitive components of scientific breakthroughs. The ontological shift proposed by Copernicus has been studied as a covariant of epistemological trends developing, initially, within the workshops of late medieval scholasticism

R. S. Cohen and M. W. Wartofsky (eds.), Epistemology, Methodology and the Social Sciences, 203–225.

including the rejection of fictional notions of hypotheses and the assertion of human impressions of nature with certainty. This experience of certainty may be related to the quest for certitude in matters of faith.[4] From this point of view, Copernicus was 'revolutionary' not merely in his rejection of particular aspects of medieval philosophy, science and theology nor only in his presentation of a model of the cosmos which had already been suggested by others, but in the significance which he attributed to human abilities to comprehend with certainty the true, physical nature and structures of the cosmos.[5] This positive evaluation of human thinking is integral to what may be called the new image of knowledge.[6] Copernicus abandons the perennial concern first expressed by Plato to "save the phenomena" or "save the appearance" by reconciling reason and experience with accepted traditional verities. At first, apologists for Copernicus try to blunt the implications of his assertions by promulgating them in the fictional terms in which medieval hypotheses were argued. However, it is not long before the new image of knowledge and the broader epistemological underpinnings for the new cosmology which it provides become explicit and inseparable from the new cosmology itself.[7] It is this shift of orientation from metaphysical speculation to the confident contemplation of nature as a more promising and certain means of knowing God's glory which 'spurs' the rapid development of modern science.[8]

In assessing the significance of this new image of knowledge in fomenting the 'deep revolution' of which Copernicus's cosmological musings are an integral part, it has been argued that in societies such as Confucian China where this image of human knowledge did not develop as part of a complex of religious motivations for investigating the divinely ordained laws of nature, notwithstanding earlier impressive accomplishments in applying human reason to the control of natural forces, scientific knowledge did not pass the threshhold it did in seventeenth century Europe.[9] An even more propitious case study of responses to the broader implications of the new cosmology can be drawn from a society, geographically and culturally contiguous with the 'European minds' allegedly affected by the new image of knowledge. European Jewry was close to the centers of the Copernican debates. A number of Jewish thinkers were not indifferent to the outcomes, though curiously unshaken – a reaction which calls for some consideration. If, as has been suggested, what was indeed 'revolutionary' about Copernicus and had an impact on the development of modern science, was the claim made to certain knowledge of the physical reality of nature, it would be interesting to measure the extent to which this claim was at issue in Jewish responses to Copernicus. Insofar as the new image of knowledge was part and parcel of the modernization of European consciousness, the discussion of Copernicus among Jews can provide us with a subtle tracer of the spread of attitudes associated with modernization.

This paper will examine Jewish reactions to Copernicus from the late sixteenth to the early nineteenth centuries. It will consider reactions to Copernicus within particular historical contexts and against the background of developments in scientific knowledge and changes in the image of knowledge, focussing on the altering bases for acceptance or rejection of Copernicus and the new cosmology.[10] But it will also attempt to take soundings of the transhistorical dialogue of self-conscious spokesmen for a tradition trying to clarify the issues involved in formulating a Jewish position on Copernicus. Finally, in comparing the reactions to Copernicus in Jewish and Christian Europe, it will question the existence of any possible religious issues which might have provided Jews with the 'spurs' for scientific investigation. It will attempt to respond to this question by reversing its very terms. Speculating upon the degree to which certain knowledge of nature through its empirical investigation could provide Jews with the certitude sought in problems of Jewish faith, including not only God's relationship to the world but the relationship of Jews to the larger society and the very meaning of Jewish historical existence, it will gauge the pressures for Jewish participation in the growth of modern science.

II

Before assessing Jewish responses to Copernicus and the new cosmology, a few more general remarks must be made in regard to the study of astronomy in Judaism.[11] Though at an early date the Jewish calendar had been fixed with great accuracy upon calculations rather than observation, awareness of its astronomical roots was preserved, arousing some theoretical interest. Interest in astrology likewise prompted astronomical investigations. Though the pertinence of these investigations to the understanding of Jewish fate was highly contested, this did not deter enthusiasts. Still another source of motivation had to do with national pride. An early Talmudic tradition in interpreting the verse in Deuteronomy (4:6) "for this is your wisdom and your understanding in the sight of people," connected it with Jewish accomplishments in astronomy (*Babylonian Talmud*, Shabbat 75a). This interpretation was later used as a statement of principle in support of the study of astronomy. However, gentile accomplishments could not be disregarded and disputes about the valence of 'Greek wisdom,' particularly where it produced claims in conflict with those of Jewish savants, had to be dealt with in discussions of astronomy as well as other sciences.[12] The universal applicability of scientific knowledge independent of its genetic origins was a conception of knowledge for which there was little support in Jewish circles until a later period. Consequently, in addition to the sources of knowledge – tradition, faith and reason – which religious epistemologies sought to reconcile, Jewish epistemology also had to consider the validity and valence of non-Jewish knowledge. At times, reason was assumed to be intrinsically harmonious with faith and traditional verities. At others it was relegated to the stock beguilements of gentile wisdom. Maimonides suggested a compromise that seemingly resolved in one formulation issues of national pride, epistemology and metaphysics. He contended that gentiles had some claim to authority in physics corresponding approximately with sublunary regions. In that the claims of the rabbis in these matters were based upon calculation and not informed by prophetic revelation, they could be contravened. However, in the study of metaphysics, including knowledge of superlunary regions, all assertions were nothing other than mere speculations. This distinction both diffused the impingement of gentile science on Judaism while it did not rule out further scientific investigation insulated from the truth claims of faith by the fictional status accorded to this investigation.[13] The boundaries between physics and metaphysics remained fluid but essentially intact as long as the distinction between sub- and superlunary regions remained salient. However, in the period of Copernicus, with the elimination of a dual-sphere universe, the boundaries between physics and metaphysics begin to disappear. In the Jewish response to Copernicus, therefore, we might search for more than speculation upon cosmic centers and the movement of celestial bodies as the broader implications of Copernicus's assertions became more inescapable.

III

An early Jewish allusion to the new cosmology appears in the writings of a younger contemporary of Copernicus living in Prague who had contact with the great astronomer's disciples. In an epistemological discussion, Rabbi Judah Loew (1529–1609), better known as the Maharal of Prague, defends but proscribes the study of astronomy as subordinate to the study of Torah, the corpus of Jewish knowledge. While echoing the distinction suggested by Maimonides between knowledge of the sublunary spheres and metaphysics, the Maharal proceeds to challenge the authority of gentile scholars even in matters of natural science in that they cannot arrive at a consensus among themselves "for each one has become wiser based on personal knowledge and insight." In this connection, the Maharal reports of one who is called the "Master of the new astronomy who presented a different diagram and he contradicted all that those who preceded him understood and presented as the figure of the path of the stars, constellations and heavenly bodies."[14] The Maharal claims that only the wise men of Israel who had traditions based on the revelation to Moses at Sinai had reliable astronomic knowledge. For the Maharal, Copernicus's assertions had no greater authority than "personal knowledge."

Another Prague Jew, David Ganz (1541–1613), makes a more direct reference to Copernicus. In the introduction to his work, *Nehmad V'naim*, he traces the development of astronomy emphasizing its Jewish origins. The astronomic opinions of earlier scholars, Ganz concedes, were not promulgated with the same authority as were their religious doctrines. Therefore insofar as more recent scientific discoveries call these opinions into question, there is no challenge to faith. Ganz further concedes that gentile preoccupation with astronomy has surpassed Jewish interest. In connection with this he notes the great progress that is taking place in astronomical research. Ganz at the conclusion of this work reports of his visit in 1600 to the observatory which Rudolph of Denmark built for Tycho Brahe and of the flurry of activities which took place there. He himself made available to Tycho Brahe the Alphonsine Tables which, he claims, a certain Jacob Alkrasi had translated from Spanish into Hebrew in 1260. In relation to recent developments in astronomy, Ganz mentions Copernicus as the greatest astronomer since Ptolemy. Ganz attributes the heliocentric model which Copernicus develops in his "marvelous, long and very deep book" to Pythagoras.[15] Nevertheless, he adheres to the geocentric model in his own astronomical writings without reconciling the differences in perspective. This reluctance at the beginning of the seventeenth century to describe the cosmos along the lines of the Copernican model is in itself not surprising.[16] The apparent contradiction with his enthusiasm for Copernicus might reflect his fictionalist approach to hypotheses. While adumbrating a theoretical defense for scientific innovations even when they conflict with traditional verities, his thinking about the new cosmology seems to be based on the old epistemology; he registers little awareness of the broader implications of that new cosmology.

What might be considered the first detailed Jewish response to and assessment of Copernicus was made by Joseph Solomon Delmedigo (1591–1655). A man of broad erudition, he was born in Crete and studied at the University in Padua where he was a student of Galileo. Subsequently he wandered throughout the Jewish world of Eastern and Western Europe settling in his later years in Turkey. For two centuries his writings provided Jews, many of whom did not learn European languages, with an important source of information about the natural sciences.

At the beginning of his work, *Sefer Elim*, he presents a series of problems exchanged between two wise men – most likely a Socratic device of the author. In the eleventh question, the existence of a new sect which negates the premises of the ancient astronomers and philosophers is reported. "Who would believe what we have heard, the earth shall tremble and quake, its pillars will be shattered," the alleged questioner states in regard to the audaciousness of the Copernican hypothesis.[17] He quickly emphasizes that Copernicus was postulating "real movements" in pressing the claims of his new cosmology. This indicates the true danger. That Copernicus was suggesting a "strange astronomy" would not distinguish his work from that of his predecessors such as Ptolemy, whose model was also based on false premises. False premises, as the logicians claim, can yield true assertions. 'For saving the appearance, astronomers would hypothesize illusory hypotheses.' From this point of view, the old astronomy did not conflict with Jewish metaphysics since statements made even about sublunary reality were not considered 'real.' By contrast, the new astronomers 'are arrogant against the ancients and they speak with haughtiness that this is the true way'. Moreover, they make claims of multiple worlds in which there may be forms of life and even people on other heavenly bodies.

What is clear from this analysis of the Copernican hypothesis is that the author of this questionary had a deep understanding of the range of problems presented by the new astronomy. Delmedigo claims that he is preparing a tract on Copernicus by the name of *Basmot*. Whether this was lost or never written is not clear. However, in *G'vurot Hashem*, a tract later published together with *Sefer Elim*, Delmedigo responds to the new cosmology with enthusiasm. He begins by demonstrating the advantages of a metaphysic based on the new cosmology over one based on Aristotelian notions of a constricted universe. Against those who think that the descriptions of the poets and the new astronomers concerning the size of the universe were exaggerated or that sayings of the rabbis in the Talmud were unfounded, Delmedigo declares with candor

> All my days I could not believe that there was no other world apart from this one . . . and that the blessed God is established upon it and is its mover. For it is not appropriate for the infinite God to contract himself and to rest upon limited matters in spite of the

fact that they may be majestic and awesome in our eyes. Before Him they are lowly for there is no one like Him, the blessed one. And when I read in the books of the [Jewish] mystics and I would see the four worlds and the Spheres in which God exalts himself all the more, my thoughts rode upon the shoulders of the philosophers, they would soar high on the wings of angels and I would rejoice . . . in that I came to recognize the extra mightiness of the creator. And joy and light were added to me, the voice of news that I heard that the most important of the researchers have begun to think in our time that the entire universe is like a lantern . . . and the candle lit in the middle of it is the solar body which stands in the middle of it and its light spreads up to the constellation of Saturn which is the last cycle of this world.

Here the implications of the new cosmology are used not only to vindicate the rabbis but to free Jewish rationalism from its Aristotelian influences. Delmedigo goes on to claim that beyond that there are other universes with their own suns. He concludes his reportage of the new cosmology with an unambiguous approbation.

And each man according to his ability and knowledge is obligated to face his maker with an offering. And if Aristotle extended God's praises in his investigation of the origins and parts of living creatures, and Galen by describing the functions of the parts of the human body, and they wanted to make known the greatness of the craftsman by the craft, how could we, a chosen people not make known among the nations the one who acts awesomely by the movements of the Big Dipper.[18]

Both the standing of the individual and the glory of the people are enhanced through the study of science. Additional support for Copernicus is adduced from his successful explanation of the parallax. In regard to this, Delmedigo claims that "no one will reject this proof other than an absolute fool and simpleton."[19] His enthusiastic defense of the new cosmology, however, is more convincing in its passion than in its pointed and systematic responses to the earlier questions raised.

Delmedigo, admittedly an idiosyncratic personality, hardly reflects major trends of thought within the Jewish community. However, his detailed assessment and reasoned acceptance of the Copernican cosmology indicate the degree to which a reconciliation with perceived notions of Jewish verities could in principle be accomplished. Like his teacher Galileo, he aroused the ire of those responsible for ideological gatekeeping though there is no evidence that it was specifically Delmedigo's cosmological musings which created conflict between him and the rabbis of Amsterdam. Nevertheless, his book was around and available to Jews whose interests extended beyond the four cubits of Jewish law. For most of the next two centuries there was little continuity to his efforts to synthesize the spirit if not the substance of Jewish mysticism and rationalism. Rationalism as a mode of religious knowledge in Jewish life remained largely moribund. However, by freeing it from "guilt by association" with the declining reputation of Aristotle, Delmedigo pointed to new possibilities.[20]

The Catholic Church, from the beginning of the seventeenth century, aligned itself with the geocentric proponents more firmly than ever before. In the effort to strengthen its ideological boundaries against direct challenges to its authority, it turned its censorial and inquisitorial machinery against advocates of the new astronomy. Among Protestants, particularly the more radical and fundamentalist, Copernicus's model was still viewed as the 'torturous interpretation' of philosophers applied to God's works, for which there is little support in the perceptions of individuals. This emphasis upon primary experience may have supported empiricist orientations developing among Protestants, particularly in England, which would lead to a greater receptivity towards the new image of knowledge, an increased interest in the study of nature and, willy-nilly, acceptance of the new astronomy.[21] By the end of the seventeenth century, with improved techniques of observation and the conversion of most professional astronomers to heliocentrism, the *Kulturkampf* was largely decided in Christian Europe. In this period during which the battle lines of cosmologies had been pitched across Europe, little response for or against Copernicus was to be found within the Jewish community.

Among several Jewish writers of the late seventeenth century whose writings were published in the eighteenth century, we find response to Copernicus. The first direct and resolutely unambiguous rejection of the Copernican model may be found in the writings of Tobias Katz (1652–1729), better known as Tobias the doctor. He authored an encyclopedic work on science, *Ma'asai Tuviah*, published in 1707, comparable to Delmedigo's *Sefer Elim*, as a basic handbook for the popularization of scientific knowledge within the Jewish community but, in its orientation, quite different. Tobias makes reference to his Jewish predecessor at the University of Padua, saying of him that "all who follow him with difficulty understand his work and people are unable to enter the inner chambers of his thoughts. Therefore I have come along."[22] This recognized and acknowledged difference in the spirit of two personalities straddling either ends of the seventeenth century calls for attention.

Similar to David Ganz, Tobias traces the Jewish interest in astronomy and the contribution of Jews to that body of knowledge. He presents a systematic comparison of the geo- and heliocentric cosmologies, attributing the latter to Pythagoras from whom Copernicus appropriated this model. Tycho Brahe's cosmological model is also mentioned, but he dismisses this as well as that of Copernicus as strange opinions. The rejection of Copernicus is explained by the assertion that "All the proofs that he and his friends brought are against the Holy Scriptures and the words of the truthful prophets who are reliable in their words." The support for the movement of the earth that "some heretics" try to evince from the rabbinic homily, which compares the Hebrew words 'rats' (run) and 'arets' (earth), is adumbrated with skepticism.[23] While avoiding a theoretical discussion of the certainty of human knowledge or the valence of gentile wisdom, he does not rest his rejection of Copernicus on the authority of scriptures. Drawing out the implications of the new cosmology, he tries to refute this model from his understanding of the experimental data of projectiles as well as the empirical evidence that people do not fall off the ground and buildings do not crumble, proving that the earth is stationary.

The difference in tone and position of these two personalities may be related to the different periods and environments in which they spent their formative years and subsequently wrote the works under examination. Delmedigo, who grew up and was educated in Italy, lived in the wake of its Renaissance and imbibed the more enduring qualities of the cosmopolitanism of Italian Jewry. Though the reactionary measures of the Counter-Reformation supported a more inner-directed orientation, Jews through the late sixteenth century maintained some reference to Christian culture. Delmedigo notes with approval and himself reflects the growing influence of Jewish mysticism, but this seems to provide for him an interpretive schema rather than an impediment to his curiosity in regard to natural phenomena. Tobias by contrast was brought up in the Germanic countries and under the influence of the more exclusive Ashkenazic culture. The ambiance which he reflects is permeated more deeply by hostile Jewish-Christian relations. More tolerant winds from the North and West had little effect upon Jewish communities, themselves plagued by heretics and ideational strife. By gleaning scientific knowledge and presenting it in a form accessible to Jews, Tobias intended, in fact, to obviate their interest and attraction to non-Jewish literature and the confusion which it might generate among Jews, rather than to broaden their intellectual horizons. The deeper inroads made through the seventeenth century by Jewish mysticism and the mythological and neo-Platonic interpretations of nature which it popularized is evident in the writings of Tobias and his contemporaries. The negative qualities which nature takes on when viewed through these lenses is illustrated in a sermon which Tobias quotes in the name of his father. Nature and creation are viewed as that which separates God from his people. This may contribute to the very different attitudes that Delmedigo and Tobias, two early Jewish 'scientists', seem to espouse towards the study of nature and contemporary trends within that study. Another element of Tobias's staunch rejection of Copernicus may be explained in relation to the scripturalism promulgated by the Protestant Reformation and its impact upon Jews.[24] In this respect there may be consistency between the approach to the study of nature and to the study of scriptures in the eschewal of "tortured interpretations", in favor of seeking the literal meaning and avowing as most real that which appears to the eye. It may be precisely because Tobias understood the broader implications of the new astronomy, and the opportunities that it would provide for its espousers to make truth claims in regard to the movement of celestial bodies, and perhaps in other matters as well against what is apparent, self-evident or reliable knowledge because of its transmission by trustworthy interpreters of the tradition, that Tobias vehemently opposes Copernicus.

The authority of scriptures as a basis for the rejection of the Copernican cosmology is argued to an even more radical degree by a contemporary of Tobias the doctor, David Neito (1654–1728), who was born in Italy, but later became the Sephardic rabbi of London. In his book *Mate Dan* or the *Second Kuzari*, styled after Judah Halevi's Socratic dialogues between a spiritually motivated king and a Jewish savant, the latter asserts that by reason alone it would be plausible to conceive of the earth as a planet similar to other planets. Moreover, he rejects the argument that astronomical knowledge is inherently uncertain as a result of the inaccessibility of the upper regions to human visitation and scrutiny at close hand. Having made this, by no means uncontroversial, allowance to the reliability of human knowledge, he proceeds to argue that assertions in regard to the rotation of the earth must be rejected because they contradict the literal meaning of scriptures. Biblical phrases which support the geo-centric model cannot be interpreted to be mere accommodations to human perception. He emphasizes that any opinion of the wise men of the nations which conflicts with the written or oral Torah must be summarily and forcefully rejected, but one that does not may or may not be accepted and becomes no more than a matter of opinion. Nieto casts his assessment of gentile knowledge in the terms of the old image of scientific knowledge with its fictionalist approach to hypotheses. At the same time he accepts the anti-Aristotelian position, most recently promulgated by Kepler, that the cosmos is expansive and includes other inhabited planets as a position harmonious with Jewish faith. In rejecting the Aristotelian cosmology, he echoes Delmedigo's argument that it adds to the glory of God to think that there are other universes and that it goes against reason to think that God created the expansive cosmos merely for the purpose of the residents of Earth. Nieto notes that, unlike in times past when Aristotle was thought to have spoken with divine inspiration, about one hundred and fifty years ago wise and perceptive men began to challenge him, asking, "Why should we believe in his opinion considering that it was wrought from the quarry of his own mind?"[25] This same man, David Nieto, subsequently created a scandal when he allegedly suggested in a sermon that there was unity between God and nature. His congregants turned for theological adjudication to the well-known hunter of heretics, the Haham Tsevi who vindicated Nieto.[26]

A younger contemporary of Tobias and David Nieto, Solomon Basilea (1680–1749) also alludes to Copernicus but like the Maharal uses it to undermine the possibilities of true and certain knowledge, and to attack Aristotle and his supporters among the Jewish rationalists. He establishes his authority in these matters by describing his broad erudition in the sciences attained when he was younger and the recognition that he was accorded by the gentiles as an astronomer. Rejecting the assertion of Jewish heretics who claim that Torah contains both grain and chaff in consequence of which Jewish positions in astronomy may be superseded by the findings of scientific investigators, Solomon Basilea reports that some of the claims made by the rabbis, particularly in astronomy, have been confirmed now that the heavens can be investigated with the aid of the telescope. Taking the counteroffensive, he asserts that astronomy is little more than the effort to explain appearances by hypotheses that may or may not be true. Much ink is spilled on metaphysical arguments but human knowledge only deals with appearances. The best and the greatest number of astronomers agree with Copernicus and some of the students of Pythagoras, but this is nothing more than opinion. Citing Plato's theory of forms, he avers that "knowledge is only available to angels who know the roots of things in the heavens from which things in the lowly world emanate." In the conclusion to his epistemological discussion, Solomon Basilea affirms the fictionalist notion of hypotheses, evoking the authority of Descartes and the tone of his skepticism.[27] The epistemological foundations upon which, from the time of Copernicus, modern science develops a century and a half later were still under the assault of a Jewish savant otherwise sympathetic to the new cosmology but aware of the corrosive effects of its epistemology on matters of faith. The resistance to Copernicus continued within the Jewish community through the eighteenth century as it did in pockets of Christian Europe. However, those who expressed opposition to

Copernicus now did so more defensively and with greater awareness of being in dissonance with accepted truth claims and common-sense knowledge of European society. Acceptance of the Copernican model was an issue among other issues within the growing polarization between traditionalists and modernizers. However, not all traditionalists rejected it nor could all modernizers be sanguine in regard to its broader implications. The growing influence of the latest elaboration of the new cosmology as presented by Newton with its mechanistic overtones and root metaphors raised new challenges to particular religious outlooks, but, as in the case of Newton himself, did not eliminate the possibilities of metaphysical and mystical investigations.[28]

Two mid-eighteenth century central European Jewish writers who maintain a flirtation with natural science, albeit with less sensitivity to intellectual currents of the larger society than their predecessors, both ultimately reject Copernicus. Jonathan Eibschütz (1695–1764), in a pietistic speech, espouses the emulation of the heavenly bodies which, as he argues, have hidden souls and rational faculties. Their regular cycles are due to their obedience to the creator. He notes the assertion of Christian scholars that stars have no independent sources of intelligence but are rather solid spheres like the earth circulating by their nature. Consequently, the rabbinic statement that the stars "move joyously to fulfill the will of their maker" evokes laughter from these savants. Eibschütz claimed that he once challenged several to a public debate on the issue of magic. At that time he argued that if they assert that stars are inanimate and move in a mechanical fashion, how can they believe, as stated in their books, that magical incantations and rituals can evoke immanent powers? Eibschütz reports that these scholars conceded to his point of the intelligence of heavenly bodies and even offered to publish his argument in Prague. In regard to the classical Talmudic text in which the Jewish scholars first claim that the constellations are fixed and the spheres move but eventually concede to the opposite position taken by gentile scholars, Eibschütz claims that both positions were in accord. For Eibschütz the new cosmology was seen as supportive of a mechanistic world-view. Against this position, he argues from Providence on the one hand and the efficacy of magic on the other.[29]

Jacob Emden (1697–1776) was a contemporary and archenemy of Eibschütz whom he suspected of subterranean affiliations with messianic movements plaguing Jewish communities of the seventeenth and eighteenth centuries. In his discussion of Copernicus, Emden calls into question the authenticity of the *Zohar*, a major text of Jewish mysticism, in which certain sources implying earthly movement were apparently being used as support by Jewish advocates of the new cosmology. He also challenges an interpretation of a verse in Psalms which may have been used to support Copernicus. However, his attitude towards Copernicus proves to be less than consistent. In his commentary to the prayer book, Emden cites approvingly the midrash rejected by Tobias as a possible proof text for the new cosmology. "There are some who comment from the meaning 'to run' as the rabbis commented . . . that it ran to fulfill the will of his maker. And there might be from this support for the opinion of the new astronomers that the earth rotates and there is no refutation from the verse, 'and the earth forever stands.' "[30] Emden here finds inadmissible the use of the literal interpretation of a verse often cited as definitive proof for the geocentric model. His discussion of models of the cosmos must be seen within the context of his more explicit concerns to affirm God's ongoing involvements in worldly affairs on the one hand and to caution in regard to the limited abilities of mortals to truly perceive the secrets of nature, on the other.[31] Against Jewish heretics whose excessive rationalistic speculations have led them to reject the belief in God's connection to the world through Torah as his revealed word and the belief in God's inclination to intervene in nature by performing miracles, Emden presents proofs even from his understanding of modern empirical science. For example, he suggests that magnetic forces and their ability to move objects at a distance "might be taken as a physical proof of divine immanence." The rejection of God's Providence by these heretics is not only impious but dangerous in that it contributes to the despair of the Jewish people. To relate world events to predetermined and unalterable mechanical forces implies that God has abandoned the lower spheres and eclipses the hopes for national redemption. Similarly, messianic activism forces the hand of God, as it were, and creates the risks that believers will be led to a sense of failure and despair. Consequently, mechanistic notions of the cosmos as well as theurgical formulae for redemption had to be firmly rejected.[32] The new astronomy was inconclusive in the support that it might lend Emden's efforts to rectify the boundaries between the natural and the supernatural, reflected perhaps in his ambivalent stance. However, his analysis points to the growing acuteness of the impingement of modern science upon Jewish consciousness and faith.

By the end of the eighteenth century, Jews interested in attaining knowledge of natural science – particularly those in western Europe who were influenced by the European Enlightenment and had mastery of European languages – could turn to non-Jewish sources. These Jews quickly accepted the latest truth claims of gentile scientists and saw the questioning of empirical science on the part of other Jews as a reflection of their backwardness and stubborn isolation from the mainstream of European culture and society. Though the polarization between those modernists who accepted the universal truth claims of science and the traditionalists who rejected it continued to grow, there were still some Jews who were concerned to construct cognitive bridges to the larger society by trying to reconcile the truth claims of Jewish traditions with the latest scientific discoveries as they were concerned to redefine the social position of the Jews within a changing European society. The discussion of the new cosmology and its implications for understanding the boundaries between the natural and the supernatural now takes place within the context of shifting social boundaries.

Eliakim Hart of London (1745–1814), somewhat literate in the scientific and philosophical literatures of his time, saw himself as a spiritual descendant of Delmedigo. Similarly and perhaps more unambiguously than his sixteenth-century master, he also drew inspiration from Kabbalah and tried to reconcile mysticism with science. Hart levels his main attacks against classical rationalists, Jews and gentiles, who "did not know of God's deeds and his creations were not perceived by the experimental methods that recent scholars have utilized. All their wisdom was in nothing but logic."[33] The discoveries of Copernicus and of Tycho Brahe and the use of new instruments like the telescope, Hart claims, prove that the ancient philosophers "knew nothing of metaphysics, and of physics and they strayed from the path of intelligence and [so] where is their wisdom and insight?"[34] He thus uses the discredited natural sciences of classical wisdom in his effort to discredit Greek philosophy and to urge Jews to abandon rationalism. He lumps together Descartes, Locke, Bayle, Spinoza, and Hobbes as atheists. He is aware of the attacks of the Deists, among whom he includes Voltaire and Hume and cites Priestley's defense of Jews whose advanced religious sentiments he took as evidence of their possession of genuine revelations.[35] Hart seeks to evoke the authority of Talmudic texts which he interprets as supportive of a posture of uncertainty in regard to the course of the stars or as suggestive of the heliocentric model. Even the locus classicus of the geocentric cosmos in Joshua (10:12) emerges as a proof text for the earth's rotation. "And he said in the sight of Israel, Sun, stand thou still upon Gibeon" is taken to mean relative to the perception of Israel.[36]

Hart's enthusiastic support of the new astronomy comes under direct attack by another late eighteenth century autodidact, Pinhas Elijah Hurvitz

(1765–1821). Hurvitz in 1797 anonymously published his *Sefer Habrit* which included an encyclopedic survey of natural science, the explicit and manifest purpose of which was to reconcile science with mysticism. He was probably not aware of Hart's work published only three years earlier in London. Hurvitz presents a double dialogue taking place between a pro- and anti-Copernican Jew and a pro- and anti-Copernican gentile. The Jewish discussants present their arguments based on scriptures; the gentiles interpret the physical evidence deduced from projectiles. He too suggests that the Copernican model is hinted at in rabbinic writings and the *Zohar*. Ultimately, he predicts, Tycho Brahe's geocentric model of the cosmos will gain universal support though he feels obliged to concede about the Copernican model that "Anyone of the people of Israel who believes in this opinion and upholds it is not thereby demonstrating any weakness of faith."[37] However, he personally questions the Copernican model. He reports experiments that he conducted involving centrifugal force and gravity. Evidently unaware of Newton's recent writings on the forces between moving bodies, he argues somewhat circularly that if the earth really moved, gravity would be unnecessary. Consequently, the existence of gravity proves that the earth must be stationary. Beyond such experimental data Hurvitz's true conviction in the matter emerges in his discussion of the image of scientific knowledge. He poignantly and pointedly argues, "And should it be that from the words of Moses, Joshua, David, Solomon and the prophets of truth we will not receive this and from Copernicus we will?"[38] In the second edition of *Sefer Habrit* published some twenty years later, Hurvitz publishes a letter which he allegedly wrote to Eliakim Hart expressing his chagrin over Hart's acceptance of the new cosmology.[39] Hurvitz himself came under attack for his astronomical musings. In an article published in *Ham'asef* in 1809, the writer, in an otherwise positive review of Hurvitz's work with enthusiastic predictions of its didactic efficacy among backward east European Jews, attacks Hurvitz for his rejection of Copernicus.[40]

As late as the first decades of the nineteenth century, the Copernican model and its epistemological underpinnings were yet to gain unequivocal acceptance within the Jewish community. Moses Sopher (1762–1830), better known as the Hatam Sopher who led the battle against modernity and modernizers in Hungary, makes reference to the new astronomy in the introduction to a tract on the Jewish calendar which he was planning to write. There he repeats the argument that the astronomic theory upon which the rabbis calculated the Jewish calendar was esoteric. Of the more conventional astronomy, the Hatam Sopher spoke disparagingly.

> Don't believe in yourself that you can understand an infinitely small aspect of it. And believe that everything with which people of our generation seek to be clever it is all vanity constructed upon the foundations of their knowledge one builds and one demolishes. Who has gone up to the heavens and descended with a spiritual bundle?[41]

Though he admits that a profound understanding of the natural intricacies upon which the calendrical calculations are based can stimulate piety, he rejects both physics and metaphysics as legitimate intellectual preoccupations.

> However we are not guarantors of this knowledge for our preoccupation is with the dialectics of Abaye and Rava (rabbis of the Talmud) . . . and our legacy is nothing more than the words of the holy rabbis from whose mouths we are sustained.[42]

It is in relation to this effort to remove astronomy and allied matters from the Jewish purview that the Hatam Sopher makes reference to Copernicus.

> And I am very surprised by a wise man of the Christian scholars who has brought a proof to this opinion that it is preferable to claim that the earth rotates about the sun . . . under any circumstance I have no evidence in my hand to refute Copernicus nor the opposing side for the proofs are formidable on both sides . . . but we don't have anything to do with the opinion of that sect.[43]

The Hatam Sopher re-appropriates the old image of knowledge based on uncertainty and fictional notions of hypotheses. He must directly confront the conflict between traditional verities on the one hand and the truth claims of science and the scientific community on the other. Yet in his advocacy of the strengthening of the social boundaries of Judaism, he not only downgrades the valence of gentile wisdom but removes the whole controversy from the purview of Jewish concern.

With the Hatam Sopher as with Milton a century and a half before there was reluctance to express preference for the old or for the new theory. In the case of Milton this reluctance stemmed from the fears of a *Paradise Lost* – the loss of certainty in regard to an ontology and spiritual geography which support as well as illustrate the truth claims of the faith and which were becoming, even in the mind of the poet, mere poetic metaphor rather than convincing assertions of what is actual and real, as a result of the growing influence of scientific knowledge. Such fears of these broader effects of the new cosmology notwithstanding, the poet could confidently indulge his curiosity about natural phenomena with some hope of a 'Paradise Regained' – that religious notions of reality would ultimately prove to be complementary to the new assertions of natural science. For the Hatam Sopher, aware as he was of the nonastronomical functions of the dual-sphere cosmology which the new cosmology eliminated, the reluctance to accept this new cosmology might be described as 'Paradise Not Surrendered:' A model of the cosmos which did not provide for a reality other than the one about which scientists were making claims of increasing knowledge of would make the worldly tribulations of the Jewish people all the more difficult to understand, and therefore to sustain. Insofar as the new cosmology eliminated distinctions between terrestial and celestial regions and claimed the universal applicability of its laws, traditionalists like the Hatam Sopher could not accept it. But to oppose knowledge of nature and the canons of scientific evidence was, for the Hatam Sopher, an unnecessary and fruitless endeavor – he could purport mere indifference.

IV

This paper has examined a range of responses to Copernicus and the new cosmology within the Jewish community over a period of more than two centuries. In relating these responses to the epistemological underpinnings of the new cosmology, we have seen how the discussion draws at times support, at others opposition from the different modes of religious knowledge of rationalism, mysticism and tradition. In addition to the growing success of the heliocentric model in gaining advocates, what changes in the European background against which Jews must accept or reject Copernicus is the more widespread conviction that scientific knowledge corresponds with empirical reality and that its laws have universal application. By the end of this period, the very basis for separate Jewish existence within European society is at least being called into question if not in actuality being altered. At the time when both internal and external pressures upon Judaism to reformulate its boundaries between the natural and the supernatural mount, pressures upon Jews to reconsider their relationship to European society increases. The confluence of these factors must be taken into consideration in gauging the extent to which Judaism might 'spur' the growth of modern science among Jews in the early modern period of Jewish history.

It bears repeating that the emphasis of this study has been upon motivations rather than institutions. The resistance to the new image of knowledge provides a partial picture of the lack of involvement of Jews in scientific investigations in the early modern period. This picture must be complemented by more information not only on how Jews were thinking but where they were in relation to main centers of scientific investigation, how they were perceived by early scientists, and the degree to which the scientific ethos was indeed universal rather than the universalization of Christian tenets of faith which might have contributed to the disinclination of Jews to participate in its elaboration. Such internal structural variables as the possibilities of the developments of new intellectual elites within the Jewish community must also be considered to complete this picture.

Here it is worth alluding again to the frequently made comparison between the development of science in Confucian China and Christian Europe. One of the aims of this paper has been to illustrate the usefulness of broadening that comparison to include Jewish Europe. The structure of the Chinese civil service which favored amateurish and classically trained savants to experts on the one hand or entrepreneurs on the other may have functioned in ways more similar to the organization of rabbinic leadership than to the academies of science developing in western Europe. In pursuing the comparison between the inhibited scientific development in Confucian China and Jewish Europe, it may be pointed out that the Judaism of the sixteenth to eighteenth centuries may have been more similar in its effect to Confucianism, which provided legitimation for the ordering of human relations, rather than to the transcendental Christianity, particularly of Puritan England, which legitimated the systematic investigation of the laws of nature.

Needham's studies of Chinese science underscore what is by now a truism in the history of science. Ontologies shape epistemologies as much as they are shaped by them – the development of science depends not only upon the accumulation of facts within the matrix of supportive institutions but also upon the ways in which people relate those facts to their conceptions of reality. If our primary focus until now has been on the inhibitions of Jews to accept the ontological shift because of their reticence in regard to certain epistemological changes, the reversal of this focus should be equally revealing. It might be argued that the resistance among Jews to the new image of knowledge had to do firstly and foremostly with a certain disinterest in nature. Certain knowledge about nature could not provide inner certitude in matters of faith for Jews who from earliest times sought to discover the hand of God in history rather than in nature. Indeed in the sixteenth to eighteenth centuries, the agonizing problem of many thoughtful Jews was not 'saving the appearance' in regard to traditional verities about the cosmos. The diffusion of mysticism as the carrier of new mythologies blunted the assaults of the findings of empirical science on matters of faith. Insofar as any conception of reality challenged Jews to surrender the 'Paradise' of an enchanted cosmos, it threatened the basic meaning structure of Jewish history. For Jews, the 'appearance' which had to be saved was that of their special relationship to God upon which was predicated the social boundaries of Jews and gentiles. Insofar as this relationship appeared to be problematic, knowledge of nature could not provide proofs of religious verities.

This assessment of different attitudes towards the relationship of inner and outer realities might be extended in viewing the interplay between ideas and historical exigencies influencing the reception of Copernicus. Among early Protestants the radical break with other-worldliness and the shift of orientation to this world as both the mirror and staging ground of salvation support a new concern for appearances. It is against this concern that the new cosmology is initially rejected as standing in violation of daily experience and the literal meaning of scriptures. But it is this same concern for this-worldly appearances that soon provides religious incentives for scientific investigation. The experiment becomes a new opportunity for the experience of revelation. A shift of emphasis takes place from the preoccupation with words about God to the concern for God's works as basic modalities of piety.[44] The confidence that one experienced in certain knowledge of God's world reinforced the inner certitude that might be experienced in regard to one's salvation. The glories of God about which the heavens now spoke and the improvement to man's estate that scientific investigations promised, to paraphrase Bacon, were consonant with the truth claims of Protestant faith.

Counter-Reformation Catholicism, by contrast, may have rejected Copernicus at first, not necessarily because it lacked the conceptual mechanism to neutralize any theological challenges of the heliocentric model but simply because it had little to benefit from the cosmological shift. The claims of the Church that it dispenses salvation, and its other-worldly orientation, would not be strengthened by the sense of its faithful having more certain knowledge of nature.

For Jews, certain knowledge attained in regard to appearances could not relieve inner problems of faith. Rational propensities were turned to the systematization of law in the great preoccupation with Torah study and even the systematization of mystical knowledge rather than the empirical investigation of nature. The problems of Jewish existence did not support a worldly focus for religious attention.[45] There was little incentive to study that nature in which Jewish history stood as such an anomaly.

It is as part of the efforts to find worldly solutions to that history from the late eighteenth century on that Jews turned to nature and its investigation. The Jewish preoccupation with words may have led, if not directly and immediately to the interest in works, at least among some early modernizers, to words about works. Historical exigencies and the new opportunities of the modern world more than the effort to see the hand of God in nature promoted the interest in modern science among Jews and their turn to scientific professions. If it no longer proved necessary for them to hear the heavens speak the glories of the Jewish God, the strong hope was retained that the situation of Jews, as well, would be improved within the estate of mankind.

Yale University

NOTES

[1] Alexandre Koyré, *From the Closed World to the Infinite Universe* (Harper and Row, New York, 1958), pp. v; 3–4; 273–76; Preserved Smith, *Origins of Modern Culture, 1543–1687* (Collier Books, New York, 1962), pp. 33–67; E. A. Burtt, *The Metaphysical Foundations of Modern Science* (Doubleday, Garden City, 1954), pp. 15–104.

[2] John Donne, 'The First Anniversary. An Anatomy of the World,' quoted in Thomas Kuhn, *The Copernican Revolution* (Harvard University Press, Cambridge Mass., 1957), p. 194.

[3] Arthur O. Lovejoy, *The Great Chain of Being* (Harvard University Press, Cambridge, Mass., 1973). Lovejoy suggests that the geocentric model was not altogether supportive of anthropocentric teleology nor flattering to the pride of man; the earth being construed more as a "squalid cellar" (pp. 102ff). Nevertheless, he finds it paradoxical that a worldly orientation rather than a compensatory otherworldliness should have developed in response to the new cosmology. "It was after the earth lost its monopoly that its inhabitants began to find their greatest interest in the general movement of terrestrial events." p. 143.

[4] See the work of Edward Grant, 'Late Medieval Thought, Copernicus and the Scientific Revolution', *Journal of the History of Ideas* **23** (1962), 197–226; 'Hypotheses in Late Medieval and Early Modern Science', *Daedalus* **91** (1962), 599–612; see also Benjamin Nelson's response, *ibid*., pp. 612–16.

[5] Edward Rosen, *Three Copernican Treatises* (Columbia University Press, New York, 1939), pp. 22–33. In connection with this, Rosen quotes Osiander's epistle to Copernicus of July 1, 1540: "I have always felt about hypotheses that they are not articles of faith but the basis of computation; so that even if they are false it does not matter . . . . It would therefore appear to be desirable for you to touch upon this matter somewhat in your introduction. For in this way you would mollify the peripatetics and theologians whose opposition you fear" (pp. 22–3). Osiander, not having succeeded in convincing Copernicus of his position that divine revelation was the only basis of truth, wrote the introduction to *De revolutionibus* which seeks to cast the Copernican hypothesis within this fictional mold. Nevertheless, from the subsequent opposition to Copernicus, it would seem that Osiander's strategy to "mollify the peripatetics" was unsuccessful.

[6] In using this concept, I follow the work of Professor Yehuda Elkana on the growth of knowledge. See 'The Problem of Knowledge', *Studium Generale* **24** (1971), 1426–1439, particularly pp. 1430–1431; also 'Scientific and Metaphysical Problems: Euler and

Kant', in R. S. Cohen and M. Wartofsky (eds.), *Boston Studies in the Philosophy of Science* (D. Reidel, Dordrecht and Boston, 1974) **XIV**, pp. 277–305, particularly p. 279.

[7] Benjamin Nelson, 'The Quest for Certitude and the Books of Scripture, Nature and Conscience', in Owen Gingerich (ed.), *The Nature of Scientific Discovery* (Smithsonian Institution Press, Washington, D.C., 1975), pp. 355–71.

[8] This argument was adumbrated in connection with seventeenth century England by Robert K. Merton in 1938. A new edition and preface of this work has been recently published. *Science, Technology and Society in Seventeenth Century England* (Fertig, New York, 1970).

[9] See Benjamin Nelson, *op. cit*. Also Joseph Needham, *The Grand Titration* (Allen and Unwin, London, 1969), pp. 14–54, particularly pp. 34–37. Needham's overemphasis on the relationship of democracy and capitalism on the growth of modern science blurs the boundaries between motivations and legitimation of institutions, a pitfall which we shall try to avoid.

[10] As the penultimate draft of this paper was being completed, the recent article of Professor André Neher, 'Copernicus in the Hebraic Literature from the Sixteenth to the Eighteenth Century', *Journal of the History of Ideas* **38** (1977), 211–226, was brought to my attention. Here is not the place to enter into a detailed discussion of his portrayal of Jewish responses to Copernicus. Suffice it to say that Professor Neher's overemphasis on the positive response to Copernicus leads him to overlook the important issues that were at stake in the conflict between cosmological models. The presentation of reactions to Copernicus as an indicator of 'religious tolerance' leads him to an anachronistic understanding of the material. It is questionable whether Jewish commentators, at least in the earlier period, experienced this controversy in terms of religious tolerance. What is certain is that the situation was far more complex than what Professor Neher would lead us to believe in such statements as "Freedom of thought was an integral part of the Jewish conception of science and the world," as this paper tries to indicate.

[11] Salo Baron, *A Social and Religious History of the Jews*, 2nd ed., Vol. VIII (Columbia University Press, New York, 1958), pp. 160ff.

[12] The classical example of this is the dispute reported in the *Babylonian Talmud*, Pesahim 94[b] on the fixity of the spheres vs. the fixity of the constellations in which it is stated that the rabbis conceded to the gentile scholars that it was indeed the constellations which were stationary and the spheres were in rotation. That the rabbis recanted in this matter was later used to suggest that the rabbis made no claims of ultimate truth in the area of astronomy as Maimonides indicates in *The Guide to the Perplexed*, Part II, Chap. 8 and Part III, Chap. 14. Still later this argument was used to indicate the admissibility of the new cosmology.

[13] *Ibid*., Part II, Chaps. 19 and 24. For an explication of the classification of science in medieval Jewish philosophy, see the articles of Harry A. Wolfson republished in Isadore Twersky and George Williams (eds.), *Studies in the History of Philosophy and Religion* (Harvard University Press, Cambridge, Mass., 1973), pp. 493–560. In Christian epistemology following Aquinas and particularly in the thirteenth century, there is a steady erosion of the authority of reason and the averroistic solution of the 'Double-truth'. This corresponds with the rise of Probabilism and the anti-deterministic trends which expand the domain of faith with the additional consequence of increasing the power of the Church. See the above cited articles of Grant. For an alternate explanation see

Maurice De Wulf, *An Introduction to Scholastic Philosophy* (Dover, New York, 1956), pp. 145–54. For an assessment of the impact of Probabilism on the rise of modern science, see Nelson's response to Grant in *Daedalus* **91**, 3. In the Jewish community, a similar decline in the authority of rationalism takes place in the wake of the Maimonidean controversies. However, the development and diffusion of Jewish mysticism as an explanatory system, as will be argued below, replaces religious skepticism and uncertainty. This is similar to the effects of the rise of empiricism in Christian Europe, particularly in those areas most affected by the decline of Church authority following the Reformation.

14 Maharal of Prague, *Netivot Olam* (London, 1961), p. 60.

15 David Ganz, *Nehmad V'naim* (Yesnitz, 1733), p. 9a.

16 Derek J. de S. Price, 'Contra-Copernicus', in Marshall Clagett, *Critical Problems in the History of Science* (University of Wisconsin Press, Madison, 1962), p. 216, indicates in regard to the period following Copernicus, during which there was no way of measuring or proving his claims, that it is "no wonder scientists remained skeptical until the new and decisive evidence was forthcoming."

17 Joseph Solomon Delmedigo, *Sefer Elim* (Odessa, 1864), p. 27.

18 *Ibid*., pp. 292, 293.

19 *Ibid*., p. 304.

20 This may be compared with the earlier attacks of Hasdai Crescas upon Aristotle. See Harry A. Wolfson, *Crescas' Critique of Aristotle: Problems of Aristotle's Physics in Jewish and Arabic Philosophy* (Harvard University Press, Cambridge, Mass., 1971), pp. 114–127. Also see Shlomo Pinas, 'Scholasticism after Thomas Aquinas and the Teachings of Hasdai Crescas and His Predecessors', *Proceedings of the Israel Academy of Sciences and Humanities* (Jerusalem, 1967), vol. 1, no. 10, pp. 1–101, particularly pp. 13–51.

21 Thomas Kuhn, *op. cit*., p. 195. Amos Funkenstein, 'The Dialectic Preparation for Scientific Revolutions: On the Role of Hypothetical Reasoning in the Emergence of Copernican Astronomy and Galilean Mechanics', in Robert Westman (ed.), *The Copernican Achievement* (University of California Press, Berkeley, Los Angeles, 1975), p. 197ff.

22 Tobias Katz, *Ma'asai Tuviah* (Venice, 1707), p. 24a.

23 *Ibid*., pp. 44a, 41a.

24 Clifford Geertz in his study *Islam Observed* (Chicago University Press, Chicago, 1975) used the concept 'scripturalism' to indicate the literalness with which scriptures are applied as part of the religious reaction to modernization. While the historical conditions under consideration are quite different than those described by Geertz, the concept is, nevertheless, apt.

25 David Nieto, *Hakuzari Hasheni* (Jerusalem, 1958), pp. 126–30.

26 *Responsa of the Haham Tsevi*, no. 18.

27 Solomon Basilea, *Sefer Emunet Hahamim* (Mantua, 1730), author's introduction, 2b–3b.

28 See Frank Manuel, *The Eighteenth Century Confronts the Gods* (Harvard University Press, Cambridge, Mass., 1959) and *The Religion of Isaac Newton* (Oxford Univ. Press, New York, 1974).

29 Jonathan Eibschütz, *Sefer Y'arot D'vash*, Part I, pp. 30b–32a.

30 Jacob Emden, *Sidur Bait Ya'akov* (Jerusalem, 1973), p. 447.

31 Jacob Emden, *Mitpahat S'farim* (L'vov, 1870), pp. 52–54.
32 Jacob Emden, *Tefilat Yesharim*, pp. 47–57, especially pp. 52–55.
33 Eliakim Hart, *Asarah Ma'amarot* (London, 1794), p. 8b.
34 *Ibid.*, p. 9b.
35 *Ibid.*, p. 10b.
36 *Ibid.*, p. 29a.
37 Pinhas Elijah Hurvitz, *Sefer Habrit* (Jerusalem, 1960), p. 98ff. Published from 2nd edition.
38 *Ibid.*, p. 101.
39 *Ibid.*, p. 102.
40 *Hama'asef* (Berlin, 1809), pp. 68–75.
41 Moses Sopher, *Kiddush Hahamah*, printed from unpublished manuscript in *Poel Hashem* (B'nai Brak, 1968), Chap. 1.
42 *Ibid.*
43 *Ibid.*, Chap. 3.
44 Yaron Ezrahi, *Words and Works in the Social Iconography of Scientific Knowledge: A Study in Science as a Cultural System* (Jerusalem, 1976).
45 The sociological implications of Jewish rational propensities as discussed by Weber, Sombart and others and the applicability of concepts such as 'worldliness' to Jewish life, particularly in the period with which this paper is concerned, requires further investigation. See Max Weber, *The Sociology of Religion* (Beacon Press, Boston, 1967), p. 246ff.

LAWRENCE SLOBODKIN

THE PECULIAR EVOLUTIONARY STRATEGY OF MAN [1]

INTRODUCTION

I shall be presenting some very simple ideas about the evolutionary process from which I shall infer that there is a general restriction on the form of adaptation of organisms relating to the timing and magnitude of environmental stresses and the pattern with which organisms minimize their impact. Humans, owing to their extensive reconstruction of sensory data, have the potentiality, sometimes realized in fact, to depart radically from the functional restrictions found in most animals.

While any organism can be thought of as responding to a sensory impression of the world, what I am referring to in man is an *introspective normative self-image* or self-awareness [2] intimately related to the various concepts of consciousness and fantasy in literary and philosophical as well as scientific contexts.

I shall be discussing areas in which my expertise is minimal. I, nevertheless, feel the problem is worth exploring from an evolutionary perspective. As discussed in Slobodkin and Rapoport (1974), responses of plants and animals to environmental perturbation tend to be such as to minimize the effects of the perturbation and are generally graduated so that minor disturbances result in relatively minor responses. It is not at all obvious how evolutionary mechanisms which, for several billion years, have been producing organisms that generally respond to perturbations in this minimal way can give rise to man, who can make very large responses to apparently minor disturbances. Certainly this is not the first such exploration, but I believe it may have some novel features.

I shall not be attempting to explain all aspects of consciousness nor of the neurological basis of behavior.

I shall first outline the general procedure that most other organisms use in their response to environmental perturbations. I shall then indicate how human behavior differs from this pattern. I shall suggest that self-awareness and self-image can perhaps account for the differences in response. Evidence for the biological nature of the property of self-image development will be presented and an evolutionary mechanism for its origin suggested. I shall

R. S. Cohen and M. W. Wartofsky (eds.), Epistemology, Methodology and the Social Sciences, 227–248.

conclude by indicating the weak points in my own arguments but I shall maintain that, despite these, the hypothesis of an innate human propensity to develop a normative introspective self-image is highly likely and is an effective counterhypothesis to hypotheses about the innate properties of man which have recently been promulgated.

SOME PROPERTIES OF EVOLUTIONARY ADAPTATION[3]

There is no evidence whatsoever that anything at all has been maximized by natural selection other than the evolutionary units' abilities to survive in their particular environment; i.e. to be adapted. Just as environments vary, so do those properties which permit survival.

In general, I can define failure in evolution as not leaving any descendants, but success can only be defined as not having yet failed. While survival is the criterion of evolutionary success, a model in which survival is maximized is unfortunately retrospective rather than predictive, unless we specify the mechanisms for ensuring survival.

I define an evolutionary unit as all those organisms that have some probability of contributing genetic material to common descendants. This is not identical with a species since an evolutionary unit's characteristics may change sufficiently with time that it will be given a different specific name than its ancestors. It is also different from a population since the members of a population may leave and join some other populations, in which case both populations would be considered part of the same evolutionary unit.

'Adaptations' are defined as those properties and responses that permit survival of evolutionary units in particular environmental circumstances. They may be anatomical, physiological or behavioral properties of the particular individual organisms of the evolutionary unit resulting from specific environmental events or they may be chromosomal or genetic properties either of individuals or of the aggregate of individuals.

Adaptations differ in their specificity. In general, more specific adaptations relate to more predictable aspects of the environment or to reliable signals of the state of the environment. Adaptations also differ in their response times. Behavioral adaptations are defined as those that are very rapid. Physiological reactions are slower than behavioral ones. Possible behavioral and physiological reactions are limited by the potentialities of anatomy and biochemistry and of course all of the properties of the organisms are directly or indirectly contingent on genetic factors. The degree to which all the aspects of adaptation are specified by the genotypes (i.e. are highly canalized),[4] as opposed to being largely determined by environmental contingency, vary from organism to organism, and from property to property within each organism. For example, the adult size of birds is very highly canalized while the adult size of many polyps is not. The behavior patterns of bees are highly canalized and the behavior patterns of man are certainly not. The iris color of man is highly canalized while the waist measurement is much more susceptible to environmental influence.

In general, living organisms are well adapted to their present environment. That is, if the means, variances, and temporal and spatial fluctuations of the various possible measurements of their environment stay constant, presently living evolutionary units will continue to survive without becoming more or less abundant on the average, and without evolving. When environmental changes occur some genetic properties that have until then been more or less deleterious, may become relatively more favorable. Environmental perturbations consist of changes in the frequencies and magnitudes of environmental properties and in the kinds of environmental problems that must be faced.

Evolution, in the sense of gene frequency change, is a last resort, implying that the physiological and behavioral devices of the organisms have been inadequate to deal with existing problems and that the genetic program underlying the phenetic response systems will be altered. In one sense, the extant behavioral, physiological and anatomical properties of organisms can be thought of as the genotypes' way of avoiding the necessity of changing.

The general response of non-learning organisms to environmental changes is a minimal one in the sense that organisms tend to preserve, as far as possible, their ability to respond normally to the unchanging portions of the environment. Any perturbation of an evolutionary unit initiates a set of responses which can be ordered in terms of the time required for their full activation. Usually, ranking responses in terms of speed of activation will also rank them in terms of the degree of commitment required by the evolutionary unit in performing each response. Rapidly activated responses generally involve smaller commitments.

The performance of a responsive act can itself act as a perturbation. Running away from a lion, for example, subjects the organism to the stresses involved in running.

The effect of this scheme is to assign the resources of the evolutionary unit in proportion to the magnitude and importance of the novel disturbance. For example, if a short-term behavioral adjustment suffices to deal with a problem tactically, and if the problem is a recurrent one, the recurring problem may act as a selective force to improve the genetic basis of the behavioral response. But if no problems of, say, physiological acclimatization are raised tactically, there is no way for the recurring problem to select directly for deep physiological changes except to a degree that the behavioral response is approximately appropriate, so that selection can meaningfully alter it.

Not only do organisms adapt to normal problems in this way and 'attempt' to adapt to novel problems in the same way, but the process of natural selection will tend to preserve the capacity to apportion responses to the magnitude of disturbances.

I have postulated a specific arrangement of responses in an optimally adapted evolutionary unit, namely, one in which responses are temporally ordered in such a way that more trivial responses have an opportunity to deal with a perturbation before the deep responses have time to reach full activation. If this scheme is correct, then qualified predictability exists in evolutionary theory, in the sense that, while we cannot predict precisely how an evolutionary unit will respond to an environmental perturbation, we can predict, in terms of the magnitude and timing of the environmental disturbance, the class of responses which will be used.

THE PECULIAR RESPONSES OF MAN

So far I have been discussing biology, some of it controversial, some of it hypothetical, but still within the area of my nominal expertise. I shall now enter the dangerous area that lies between the biological and social sciences.

If the earlier analysis is even approximately correct, then human behavior seems peculiar. People obviously respond to environmental perturbations. They shiver when they are cold, seek water when thirsty and urinate when their bladders are full, just as do other mammals. In addition, however, they may respond with extravagantly large-scale behavior to what would seem rather trivial stimuli from a purely biological standpoint. They will bomb each other because someone spit on a flag, immolate themselves or publicly chop off their fingers to emphasize that their political philosophy diverges from someone else's, accept martyrdom over the wording of a credo and so on. These typically human peculiarities are of course the central themes of art, literature, moral philosophy and religion and it is *chutzpah* of a remarkable kind to attempt to deal with them in an evolutionary context. I nevertheless enter this area, for two reasons.

First: How the pattern of minimal response to perturbation, discussed above, can evolutionarily lead to an organism capable of such massive responses to seemingly minor disturbances constitutes an interesting problem.

Second: It seems that the political and social dangers of dealing with human behavior in the light of false understandings of evolution, or no understanding of evolution, may be very great.

There seems to be a clear agreement among modern anthropologists that human activity is a complex resultant of the interplay between functional responses to environmental circumstances, sensed goals, social preconditions and history. There is general agreement that the properties of culture cannot depart too far, or too long, from the criteria that would be set by biological standards of adaptation, but I do not find clear unanimity on whether or not culturally determined standards of behavior always originated for adaptive reasons. Some anthropologists seem reluctant to admit of non-adaptive properties of culture, in the sense that if a particular cultural component can be interpreted as adaptive, they will do their best to see it that way. This may be a result of coming from a culture in which adaptive behavior is seen as virtuous and in which the plausibility of human behavior is assumed. Wittgenstein noted, with regard to the curious behaviors reported in Frazer's *Golden Bough*: "Nie wird es aber plausibel, dass die Menschen aus purer Dummheit all das tun!"[5]

Quite often surprising activities can be shown to be adaptive in fairly subtle ways. For example, when hunting is poor the Indians of Labrador use divination by scapulomancy to determine the next direction to take in the hunt. The cracks that appear on the roasted scapula are used as a route map. The scapula in this case is acting as a random direction indicator, which may be the best procedure when normal decision procedures have failed.[6]

The elaborate pig-sweet potato-warfare interactions of the Maring of New Guinea have been viewed by Rappaport as a way of maintaining an ecological steady state. Simplifying somewhat, the system works as follows. A man owns pigs and he and his wife have a garden in which they raise their starch crop to feed themselves, their children and, to a degree, the pigs. If everything prospers, the number of pigs and children increases and the fraction of the garden produce used by pigs also increases. After several years the pig herd has become a major drain on the garden, creating a necessity for more land. If this cannot be acquired by peaceful means from relatives, war with the neighbors becomes necessity. To prepare for war a feast to both friendly gods and human allies must be given. The gods consume the odor of roast pig, while the flesh is consumed by humans. The war can now be fought – which is usually relatively minor as wars go. The handful of casualties in the short war may in fact help stabilize the abundance of humans. The killing of the pigs has already eased the problem of an excess pig population and has provided a major protein source for a fair number of people. Some land re-allocation may have occurred and the shortage of pigs for a war feast means that no new war can be started by the same man for several years – until his pig herd increases again.[7]

Despite the simplifications of this brief abstract, notice that while a simple integration over these events does actually act to maintain the ecological steady state it seems like a wonderfully elaborate way of doing so. There is a persistent possibility that the supposed adaptive value of at least some of the behavior reported in the anthropological accounts is an artifact of the anthropologists' conviction that behavior 'ought to be' adaptive.

In most examples of this kind, the adaptive activity is embedded in a context of more or less explicit social mythology so that an answer given by a participant to the question, "why do you do that?," will often contain more than a utilitarian assertion but will be part of a myth or story explaining social properties of a general kind. This myth itself is treated as something of value by the society. Action will be taken in its defense.

Most organisms respond to their environment more or less directly, and such response is likely to be adaptive. Men also respond to their environment, but their response is, at least sometimes, related to a more or less explicit story about the world. For most organisms the magnitude of response to a disturbance is proportional to the importance of the disturbance in disrupting the adaptive state of the organism. For men the response is proportional to the degree that it disrupts the *perception* of propriety and well-being.

It may be validly argued that this is not uniquely human since any organism that can learn is building an introspective world of some kind and will respond in the relation between a particular event and its 'knowledge' of the world. I suspect, however, that the problem has somewhat deeper implications.

A 'normal' (i.e. obviously non-human) animal capable of learning will 'assign' to certain events the status of signals, which will be responded to in terms of their meaning rather than as simple stimuli. For such an animal the set of signals comprises a changing image of the world, to which he responds. A man feels himself to be doing more than that. He feels himself to be part of a scenario in which there may be a goal, a reward (precisely what does not exist in normal evolutionary adaptation), or at least a cohesiveness of plot and for the sake of that goal or reward or for the maintaining of the consistency of that plot, all men are capable of behaving in a thoroughly non-adaptive way on occasion.

Wagley[8] has pointed out that the Tapirapé tribe of Brazilian Indians had a profoundly pessimistic attitude towards population control. They believed that since food and other supplies are certain to become short, it is very important to have as few children as possible. The closely related Tenetehara are more optimistic and believe that if you have plenty of children, then something will happen to permit you to take care of them. The contact with the more modern Brazilians has introduced new sources of mortality into both of the Indian tribes. The Tapirapé were almost destroyed by insisting on their idea of a crowded universe in which extra children could not be supported even though new mortality sources were in fact creating space in that universe. Their population declined to 40 or 50 individuals from several thousand. This clinging to action related to goals or world images may be called heroism, bigotry, orthodoxy, etc., but is clearly something most animals do not do, or at least do not do for very long.

It may be argued that clinging to learned behavior patterns in inappropriate circumstances may be found in most mammals. Only in humans, chimpanzees and orangutans, however, has it been demonstrated that behavior can be activated by the organism's image of itself.

I believe that I am restating, in a slightly different context, one of the general assumptions of anthropology, sociology and psychology. Consider the following quotations:

> Only man is capable of asking the question, what am I?

> . . . humans may fight defensively or may be led to believe they are doing so. Animals cannot be fooled because they have no culture. Humans can be led astray by the very symbolic process which is their major adaptive tool.

> Identity of self appears always to have been a human need.[9]

> . . . awareness of the self as an object acts as a feedback system which forces the individual to alter aspects of himself in the direction of his conception of what a correct person should be. When self-consciousness is viewed from this perspective, the person's awareness of himself must figure into any ongoing behavior. The self-conscious person responds not only to external stimuli, but also to himself as a stimulus.[10]

While the content of culture may be remarkably various and determined by an elaborate combination of circumstances, the ability to participate in culture generally requires certain genetic potentialities. There are definite features of neural, muscular and skeletal anatomy needed for speech. These may be highly canalized while the content, and the form of speech is not. Analogous remarks apply to the ability to use tools.

Since speech and tool-making are much more strongly developed in man than in other organisms, we consider speech and tool-making as prerequisites or corequisites of culture. To avoid problems of the precedence of chickens and eggs, I would rather consider the genetic infrastructure of speech, tool-making, etc., as corequisites of culture.

If we now find some other human property which is as universal to all cultures as speech and tool use and that is not found in as well-developed a state among other animals, can we, at least as a working hypothesis, consider it to have a genetic infrastructure (in the same sense as speech and toolmaking) even though we cannot designate a specific anatomical basis for the property in question?

Learning ability seems to be one such property, but I don't think that the ability to learn in a general sense will permit man to depart from the normal pattern of evolutionary strategy by over-responding to small-scale perturbations – unless the environment changes so that learned responses change their significance. Humans, however, can change their interpretation of the meaning of a signal even when no environmental change has occurred. On occasion this can be a very fruitful thing to do but it can also, on occasion, be disastrous. I suggest that the feature of man that has permitted deviation from standard evolutionary strategy is that component or epiphenomenon of general intelligence which has variously been called consciousness, self-awareness or self-image. This intellectual property is an indispensable part of the equipment needed to build theories about the world or to make purposive plans whose execution involves at least temporary deviation from highly adapted states.

I focus on self-awareness rather than language for three reasons.

First: purposive, planned activity involving interactions between organisms seem possible without language. Consider the play activities of children from three to five who do not share a common language, or the movements of animal groups.

Second: despite the recent experiments in teaching language to chimpanzees,[11] the free use of language seems to be confined to humans while, clear evidence of self-awareness is available in some apes, indicating an evolutionary primacy of self-awareness over language.

Third: It seems dubious that language could develop in the absence of self-awareness. If language develops in the absence of self-awareness, what is there to say? But once self-awareness has developed, linguistic communication takes on tremendous importance.

In short, I assert that men have a normative introspective self-image, portions of which may be shared to produce a culture-specific world image, and that response to environmental events is to a variable extent filtered through the impact on these images. This image and responses to it may be adaptive but need not be. This is certainly not original as an observation.[12] It is subjectively completely obvious that at least I and the reader have such an image.

I am not asserting anything about the content of the self-image nor am I specifying how articulately its possessor can describe it, simply that all men have one. If the image interacts with the environment in such a way as to cause obviously maladapted behavior, then several possibilities are available. Either the image is altered to make adaptive behavior fit into it (which is referred to as a personality alteration on an individual level, or acculturation on a group level) or else the image does not change and the people involved die or become miserable. Which of these events actually occurs depends, at least in part, on the environment and the content of the image itself, as well as on the details of social context.

BIOLOGICAL EVOLUTION OF NORMATIVE INTROSPECTIVE SELF-IMAGE

So far I have listed biological and anthropological truisms. The next statement is a hypothesis which generates a series of less obvious questions. Some are evolutionary, while others relate to such problems as human psychology, psychiatry, linguistics and religious history. I will be concerned here only with the biological. I assert the hypothesis:

The human capacity for having a normative-introspective self-image is a biological rather than a cultural property, although the content of that sef-image is culturally determined.

When I say that an organism 'has' a normative introspective self-image, there are three assertions being made:

1. The sensory and neurological machinery of that organism is such that in the process of learning about the world, the organism will learn about itself as an object in the world.

2. The specific self-image of each organism will be contingent on its particular history as modulated by its sensory and neurological competence.

3. If the environment is such that it does not provide suitable sensory data

for the construction of a reasonably consistent self-image, the organism will be behaviorally and, possibly, physically incapacitated.

It should be noted that not all organisms have the machinery referred to in the first assertion, so that there is an evolutionary problem in defining the conditions and prerequisites for its development. These assertions are not empty since behavioral experiments, whose results can be most parsimoniously explained in terms of self-image, have been performed.

There is ample psychological evidence for the existence of a normative self-image in man, which I need not review. My concern is the evolution of the property.

If this is in fact a biological property, then I must be able to show how normal natural selection could have produced it. A general assumption of evolutionary theory is that all the intermediate stages in the development of any biological property whatsoever must have been adaptively advantageous, although not necessarily in the same way. Discussions of the origin of flight in birds, for example, infer a selective advantage to gliding behavior before true flight developed. In the development of flight in insects, however, gliding behavior may not have been significant, since the body of insects is so small that it tends not to be damaged by falling. An early suggestion was that the lateral extensions of the thorax, which were the primordial insect wings, may have developed as stabilizers in water.

Therefore I must be concerned not only with the adaptive significance of self-image but also with the adaptive significance of the various preconditions that permit the evolution of self-image.

Also, if the development of a self-image in man is a biological property, either it is unique to man himself or it is unique to the phylogenetic line that led only to *Homo sapiens*, or it is found in animals other than man.

Biologists are reluctant to ascribe biological properties to single species, unless they are of a rather trivial kind. The general feeling is that the properties that show up as highly developed complex structure in one species are found in some, perhaps rudimentary, state in related species and were originated in some ancestral form that may or may not now be determinable. As a biologist I would therefore be happier if I knew the self-image property to exist in non-humans also. If we are forced by the facts to make the assertion that it is unique to man, we must do so, but my own prejudices are against that necessity.

There is however evidence for something very similar to what we are discussing being present at least in the higher primates and perhaps in most social mammals.

We are not concerned here with intelligence as such but rather with a property more related to personality than to I. Q. If an organism possesses this property we would expect it to show signs of actively maintaining its own concept of itself.

Gallup[13] performed the following fascinating experiment. He provided chimpanzees with mirrors near their cages. They responded initially to the mirror as to a strange animal and then after three days began to use the mirror to examine parts of themselves they would not otherwise see – i.e., the way people use a mirror. When they had become accustomed to the mirror, they were anaesthetized and their eyebrow and the top of the opposite ear were painted red with a non-greasy odorless dye. On waking they were observed in the absence of the mirror and essentially no motions touching the dyed area were made. After the mirrors were replaced, the chimpanzees immediately tried to rub off the dye.

As a control, naive chimpanzees were painted and then presented with the mirror. They made no attempt to rub off the dye.

The chimps had learned to have a visual self-image during the period of exposure to the mirrors. Gallup concluded that he had found the "first experimental demonstration of the self concept in a sub-human form" and since similar studies on macaques gave negative results, he suggested that "the capacity for visual self-recognition may not extend below man and the great apes."

Lethmate and Dücker[14] have provided mirrors to orangutans, chimpanzees, two species of gibbons, four species of macaques, mandrills, Hamadryad baboons and two species of monkeys in zoos. Of these only the orangutans and chimpanzees showed self-directed image behavior including such things as "trying on" bowls as hats in front of the mirror.

It is possible that normative self-images may be fairly widespread in social animals but may not be related to visual imagery. In fact, any owner of a cat or dog will have a strong sense that this is so, and that the behavior of an animal is characterizable by its level of self-concern as much as by its puzzle-solving ability. It seems possible that odor-related self-images may be very widespread but no clear experiments have been done in this area.[15]

There is a paucity of explicit experimentation on self-image, in contrast to the great mass of material on intelligence, trainability and so on. The experiments may have to be ingenious but they seem worth doing. There is, however, a great deal of observation that is amenable to interpretation in terms of self-awareness or self-image, without being critical tests of their occurrence.

For example, there is circumstantial evidence from macaques, despite the fact that macaques' failed the mirror test. Provisioned free-roaming troops of Japanese macaques have been observed for more than 20 years and on several occasions innovative behavior has occurred. This has involved, among other things, washing of sandy sweet potatoes, seasoning of sweet potatoes with salt water, the separation of wheat from sand by floating handfuls of the mixture, and learning to swim.[16] These events, while fascinating in themselves, do not bear any immediate relevance to our problem. What is of possible relevance, however, is the fact that for each one of these innovations there was a failure of immediate and general acceptance of the apparently obviously advantageous behavior by other members of the troop. It is also of interest that (with the possible exception of wheat washing) the animals that more readily accepted the innovations were younger rather than older.

This can be explained in a variety of ways: for example, it is possible that innovative behavior declines with age and that imitative behavior is greatest between animals of comparable ages. One of the possible explanations, however, is that older animals tend to have a more clearly defined introspective self-image which makes them relatively immune to behavioral changes. This suggests possible experiments in which new behavior is taught to one or a few animals in a social group and then the response of the untaught animals to that behavior is observed. That is, it may be possible to delineate the self-image of an organism in terms of seeing what behavior gives offense to that self-image.

Duval and Wicklund[2] in their study of "objective self-awareness" (i.e., consideration of the properties of the self as an object by humans) have collected a great deal of information on the tendency of humans to alter their own behavior so as to conform to the behavioral decision norms of a group and have related this to self-awareness. Kummer and also Rowell describe the events preceding a baboon troop's choosing a direction for movement after a night's rest.[17] Single males start in a particular direction for movement but unless they are followed by a reasonably large number of animals they return to the main group and sit down. Typically several abortive starts are made by different males. Only when a male is followed by several of the high status females does the entire troop join in following him. His 'choice' of direction, and the direction taken by the troop 'following' him is therefore a group concensus. A possible, but not necessary, interpretation of this behavior might be in terms of the male's readjustment of his own image of the morning's route to conform to that of the other members of the troop.

There is a fair amount of similar observational data which supports the idea of normative introspective self-image in the same curious way that the various sacred kings listed by Frazer support his explanation of the events at the oak grove of Nenni. I shall not recount these observations here. However, from the mirror studies it is clear that normative self-image is not confined to humans. It should therefore be possible to consider how this property may have evolved. I can attempt to define the sort of environmental problems which would call for the development of self-image as at least part of a possible solution. I should also be able to specify the biological prerequisites that would make this one of the class of probable solutions.

It might be helpful at this point to return and reconsider the peculiar nature of evolutionary predictions. There are reasonably stringent restrictions on the class of responses evolutionary units will use, given novel selective pressure, but in general it is not possible to predict precisely how a particular novel selective force will alter the anatomical, behavioral or genetic properties of a particular evolutionary unit. If mean selective effects on particular genetic alleles are known *a priori*, stochastic predictions of future allelic frequencies can be made, but assuming *a priori* knowledge of selective forces begs the issue of prediction.

The impossibility of either stochastic or deterministic prediction is due to the fact that in most circumstances, for most evolutionary units, there exist empirically distinguishable responses to selective force which meet equally well the known restriction on the acceptable family of responses, even if these are well defined. The underlying basis of this multiplicity of functionally equivalent responses is the fact that ecological systems in general, and evolutionary units in particular, are complex feedback systems in the sense that particular properties of the system may simultaneously act as rate regulators for certain other properties of the system, as null setting for other feedback processes and as correctors of deviations in yet other processes, all the while they are being regulated themselves. In this type of system what has been called 'cybernetic causality' holds – that is, there are certain forbidden states the system will not reach under a given perturbation and these are, in principle at least, clearly definable. It may even be possible, at some time in the future, to assign a probability to each of the clearly definable distinct non-forbidden states and probability distributions over those that are not distinct. However, the system cannot be uniquely assigned to only one of the permitted states.

For present purposes it is of interest to define the class of far-fetched or inadmissible evolutionary explanations. These are explanations which violate the accepted restrictions on evolutionary process. For example, evolution is understood to be non-teleological, so that any explanation which required some characteristic to develop up to a useful state through a series of useless stages would not be acceptable.

Further, evolutionary changes which benefit the perpetuation of genetic material in other individuals of the evolutionary unit more than the genetic material in their possessor are purely altruistic and are therefore not acceptable. Also, explanations which require massive changes in an evolutionary unit are *a priori* less acceptable than explanations that require only small changes. This implies that in discussing any evolutionary change in response to a particular environmental condition it is of major importance to specify the properties of the organisms themselves prior to the onset of the change in question. For example, various organisms have been faced with the problem of rapidly moving through forests. The texture of the forest, and the degree of rapidity required, place certain restrictions on the class of acceptable solutions – but the preadaptations of the organisms themselves also set up restrictions. Small primates and squirrels use jumping and holding, birds fly, gibbons brachiate (and I have been told that hyraxes simply drop to the ground).

Explanations that require curious and unlikely co-occurrences of environmental events share the low likelihood of those co-occurrences. Conversely, an explanation that avoids these and similar, implausibilities must be accepted as plausible – without any commitment to the plausibility ranking of other, as yet unspecified, explanations.

If we are permitted to assume ancestral primates or preprimates with familial ties so that several generations live together in a band or troop and with strong ability to learn and remember, our problem is relatively easy. We need only further postulate an environment in which serious problems, capable of behavioral solution, arise with a frequency of approximately once to twice per generation. In fact many large mammals and in particular many primates meet these requirements. Then the statement by Rowell,[18] writing about olive baboons in Uganda, is of dramatic relevance:

> In a long-lived group living species, an animal increases in value to its group the more experience it has accumulated. For example, exceptionally bad droughts tend to occur at intervals of around 20 years. Groups with individuals old enough to remember how the last disaster was survived will obviously be at an advantage in being led to remaining food and water supplies and it will probably be the aged individuals' own descendants who are so protected. Thus there will be a selection pressure for longevity as such even beyond reproductive age. Presumably, more selection for longevity in females if the males are less likely to remain with their own offspring.

As a final speculation, consider what is involved in an old animal remembering for 20 years what occurred in a bad drought. Under most normal considerations of learning theory an organism is assumed to require reinforcement, or at least review, to prevent it from forgetting what was learned a long time ago. A drought may require discontinuous changes in behavior; for example, going to a completely different water hole rather than the one that you have usually gone to. If the drought water hole is simply not used during non-drought periods, how does a baboon retain the information gained at the age of one or two years so that it may be used 20 to 30 years later? If the animal builds the information somehow into his own personality, and reviews it internally, we expect that it is not likely to be forgotten. Also the degree that the animal is capable of learning new information is likely to be diminished somewhat, since old mammals may be less teachable than young ones.[19] You can't teach old dogs new tricks, unless you do the teaching through a careful consideration of the personality properties of the particular animal involved and this is again commonplace information to anyone who has owned a pet for several years.

At this point the existence of self-awareness in animals other than humans has been demonstrated. It has been suggested that the capacity for self-image requires a social nexus as an evolutionary prerequisite, while it, in turn, serves as a corequisite for the development of linguistic capacity. In addition one plausible suggestion has been made as to how the capacity for self-image might have developed as a consequence of natural selection.

While the details of the evolution of the human capacity for self-awareness are not in fact known, there is no indication of anything uniquely difficult in this given the prerequisites of long life, learning ability and social interaction.

The evolution of long life, social interaction and learning ability do not themselves require any violations of normal evolutionary considerations and are in fact intimately interrelated.[20]

SUMMARY, CONCLUSIONS AND APOLOGIA

I have discussed the human tendency to develop a normative introspective self-image. My attention was focused on this problem by noting that what I had hoped to be a general theory of optimal evolutionary strategy did not adequately account for the development of human behavior. Self-image permits a series of deviations from the pattern of evolutionary strategy found in most animals.

Normal evolutionary strategy consists of making minimal responses to environmental changes. This forces a kind of adaptive continuity on evolution. That is, an evolutionary unit cannot reach any particular adaptive state except by a path which maximizes adaptation along its entire length. By contrast, a 'conscious' individual organism, or (to the degree images are shared between organisms) an evolutionary unit, can deviate from the path of local maximization of adaptation 'in order' to reach some 'desired' state, which may or may not be a more highly adapted or more advantageous state than it could have reached by the normal path.

Teleological behavior, in which a group of organisms enter highly maladaptive situations 'in order' to reach some goal seems to depend on either all of the organisms, or at least their leader, having some image of the goal. This is not the same as claiming that all apparently teleological behavior requires an internal image. In climbing a hill, for example, a behavioral response of moving higher will bring an organism to the top of the hill, if the hill is monotonic, without its ever having had an internal image of getting to the top.

While introspective normative self-image has been demonstrated in chimpanzees, orangutans, and man, the relations between self-image and 'motivation' or 'purposiveness' have certainly not been completely analyzed. It is possible, for example, that self-image and motivated behavior will be found to be coextensive – but this will certainly depend on the precise definitions used.

The hypothesis that humans construct a normative introspective self-image whose content is not genetically determined has the extremely important implication that human fantasy is a major element in human behavior. There is no extant evidence that there is any constraint on that fantasy from 'within,' however environmental circumstances might reward or punish the actions derived from fantasy. The force of this assertion is that a man can develop a self-image in which he considers himself to be purely the result of operant conditioning, or in which he considers himself to be naturally aggressive, constrained only by environmental forces, or naturally loving although occasionally required to perform aggressive acts by environmental forces. In short, a self-image can be constructed in the mode of Skinner or Ardrey or Ashley-Montagu but the only biological given need be the tendency to build a fantasy about one's self.[21]

There may be images which cannot possibly arise except in particular environments. There may in fact exist a class of self-images which are structurally impossible, in the sense of being unable to be fantasized by the human brain. The investigations by the Lévi-Strauss school of anthropology, which examines extant world images, may eventually show that the set of images is closed under certain transformation rules, which are themselves based on neuroanatomy and physiology.[22] In the same sense the Chomsky hypothesis of innate grammar,[23] if verified, may show that the linguistic mechanisms for communicating fantasy, and thereby, for building the structures examined by the structural anthropologists, are also under biological limitation. As of the moment, however, there is not, to my knowledge, any visible or definable border of this sort to the set of possible self-images.

The entire area is sufficiently unexplored and so far from the normal considerations of a biologist that one obviously questions, why enter it at all? Notice that the portion of the above discussion that deals with evolutionary strategy is a general theory, based on elementary principles and presumably of wide applicability. The discussion of the properties of the human and the arguments designed to fit human properties into the general scheme are largely *ad hoc* – an attempt to save the general theory in the face of the challenge provided by man.

Evolution of the capacities for culture – in particular the capacity for self-image development – presents a serious evolutionary problem, since an organism with this capacity has such remarkably poor canalization of its behavioral development as to obscure the effects of natural selection on specific behaviors.

The social sciences take as their subject matter a humanity equipped with all the biological prerequisites to culture. It seems to be a biologist's problem to demonstrate, at the very least, the plausibility of the evolutionary development of these capacities.

Obviously I have made neither a definitive nor complete analysis of this problem. I hope that I at least may help focus interest on it.

In writing this material at the present time it would seem that I have violated Wittgenstein's dictum, "Whereof one cannot speak, thereof one must be silent!"[24] But, there are considerations that make it not only permissible, but even urgent, to enter this difficult area. Wittgenstein was apparently attempting to discourage scientists, mathematicians and philosophers from making assertions which could not be properly formulated within the limits of scientific language. Most particularly his dictum prevents relegation of moral responsibility to reified pseudo-scientific assertions, since ethical dicta would not be well formulated in his sense.

Practically, the terminal assertion of the *Tractatus* forbids statements within science about the ultimate nature of anything at all, in particular statements about the ultimate nature of man.

Unfortunately, the history of the twentieth century has made it abundantly clear that pseudo-scientific assertions about the nature of man still have powerful political meaning. Even if their promulgators are politically ignorant, innocent or naive – as they often are – the fact that the assertions have been made by persons in the official role of scientist, or scientific reporter, permits persons and agencies with particular self-serving concerns to deny the moral responsibilities for the consequences of their decisions in the name of science.[26]

Science in the twentieth century has for the general public the same role of justifying political action that organized religion and philosophy have had in the past, despite repeated rejections of this role by competent scientists.

There has recently been a plethora of statements of the form: "It is the biological nature of man to *x*," where *x* is a statement specifying a set of activities or attitudes which men are supposedly constrained, in some sense, to perform or to 'have' by the 'laws of nature.' In just the last twenty years Ardrey, Lorenz, Turnbull, Mead, Skinner, Wilson, Alexander, Collinveaux, Hardin, Ashley-Montagu, and I, and a host of others have made statements of this general form. Refutations of most of these assertions can be made – taking them one by one – but the refutations will not have the force of the original assertions for several reasons.

To properly deny that, for example, "Man is naturally aggressive," is not equivalent to the strong assertions of negation that "Man is naturally *not* aggressive," nor even to the assertion, "Man is not naturally aggressive" but rather to the assertion:

"The statement, 'Man is naturally aggressive,' is a nonsense statement which should not be taken as scientifically valid or even meaningful."

By declaring an assertion to be nonsense we are leaving the lay audience with an uncomfortable void in their attempt to answer to themselves their own questions about the "nature" of man. To assert that these questions also are meaningless is not pragmatically acceptable because it denies the role of science as answerer to all possible questions – leaving the field free for even more dangerous sources of answers.

Rather than simply denying the various current assertions about man's innate moral qualities and behavioral propensities I make the positive assertion that it is in the nature of man to build mental images of himself – as well as of other things in the world – and that the properties of his personality are to a tremendous degree products of his fantasy as it interacts with his environment. Man imagines his own properties and then conforms to his own imagined properties – violating the theory of types completely. If this generates paradox, then man lives in paradox. Paradox presents a problem to the logician in his working hours, but perhaps not in the rest of his life.

Granted, the hypothesis is neither completely analyzed nor as well documented as I would like. It is, however, certainly better founded than any of the restrictive assertions about man's nature that are now loose in the world. It also takes account of the enormous richness and diversity of articulated human aspirations and is considerably more sanitary from both a political and philosophical standpoint than alternative arguments.[27]

State University of New York at Stony Brook and
Smithsonian Institution, Washington, D.C.

NOTES

[1] I am happy to acknowledge that this was written while a Guggenheim Fellow. The biological theory was developed with the aid of an NSF Grant in General Ecology. I have benefited from criticisms by D. Futuyma, R. Alexander, G. G. Gallup, G. E. Hutchinson, Katherine Ralls and my wife, Tamara Slobodkin, among others. Some agree with me, some do not.

[2] The term 'self-awareness' is used in essentially the sense of G. G. Gallup, Jr., 'Toward an operational definition of self-awareness', in R. H. Tuttle (ed.), *IXth International Congress of Anthropological and Ethnological Sciences. Primatology session, subsection D.* (Mouton, The Hague, 1976); and of S. Duval and R. A. Wicklund, *A Theory of Objective Self Awareness* (Academic Press, New York, 1972).

[3] This section is mainly from: L. B. Slobodkin and A. Rapoport, 'An optimal strategy of evolution', *Quart. Rev. Biol.* **49** (1974), 181–199 in which the rationale for many of these statements can be found.

[4] The idea of 'canalization' has been developed by C. H. Waddington, *The Strategy of the Genes* (Allen and Unwin, London, 1957). A highly canalized property is one which develops in much the same way regardless of environmental variation during development. Poorly canalized properties are not developmentally restricted in this way.

[5] Wittgenstein, L., 'Bemerkungen über Frazers *The Golden Bough*', *Synthese* **17** (1967), 223–253.

[6] Moore, O. K., 'Divination, a new perspective', *American Anthropologist* **59** (1957), 69–74.

[7] Rappaport, R., *Pigs for the Ancestors* (Yale University Press, New Haven, 1968).

[8] Wagley, C., 'Cultural influences on a population: a comparison of two Tupi Tribes', *Revista do Museu Paulista*, new series, **5** (1951), 95–104.

[9] This and the preceding two quotations are taken from Alland, A., *The Human Imperative* (Columbia University Press, New York, 1972). Alland carefully refutes many of the assertions of genetically fixed human nature.

[10] See Duval and Wicklund (note 2 above).

[11] See, for example, the collection of papers in *Behavior of Non-Human Primates: Modern Research Trends.* Vol. 4, A. Schrier and F. Stollnitz (eds.) (Academic Press, New York, 1971).

[12] The early sociologists focused on the problem of the development of the self as absolutely central to the development of social life. Consider for example: Mead, G. H., in C. W. Morris (ed.), *Mind, Self and Society from the Standpoint of a Social Behaviorist* (University of Chicago Press, Chicago, 1934); Cooley, C. H., *Human Nature and the Social Order* (Shocken Books, Inc., New York, 1964); Merleau-Ponty, *The Primacy of Perception* (Northwestern University Press, Evanston, 1964). This type of sociology seems to have been largely eclipsed by the more convenient and less profound concerns with the massive manipulation of dubious data by elaborate statistical machinery. Some of the weaknesses of the sociological lust after the quantitative are discussed by Deutscher, I., *What We Say/ What We Do* (Scott Foresman and Co., Illinois, 1973).

[13] Gallup, G. G., 'Chimpanzees: Self-Recognition', *Science* **167** (1969), 86–87. (See also note 2 above.)

[14] Lethmate, V. J. and G. Dücker, 'Untersuchungen zum Selbsterkennen im Spiegel bei Orang-utans und einigen anderen Affenarten', *Zeitschrift für Tierpsychologie* **33** (1973), 248–269. The species examined by them were *Pongo pygmaeus, Pan troglodytes, Hylobates lar, H. agilis, Cebus apella, Ateles spec, Papio hamadryes, Mandrillus sphinx* and *Macaca silenus*. Gallup also examined *Macacus arctoides, M. mulatta* and *M. facicularis* with color markings. Lethmate and Dücker used color marking only on *Cebus apella* and on the chimpanzees and orangutans. Only *Pongo pygmaeus* and *Pan troglodytes* showed self-awareness – all others responding to the mirror as to another individual.

[15] This was suggested in a conversation with John and Devra Eisenberg at the National Zoo.

[16] J. Itani and A. Nishimara have recently published an extensive review of the famous Japanese monkey observations. [The study of infrahuman culture in Japan, a review symposium of the Fourth International Congress of Primatology, Vol. 1, E. W. Menzel, Jr. (ed.), pp. 26–50 (Karger, Basel, 1973).] They refer to a long series of papers by Kinju Imanishi in which the psychoanalytic concept of 'identification' is used to explain the history of individual monkeys. Specifically, the 'personality' of individual monkeys is seen as developing by animals imitating specific elders. Implicit in this seems to be at least a concept of subject self-awareness, if not normative introspective self-image.

[17] Kummer, H., *Primate Societies: Group Techniques of Ecological Adaptation* (Aldine-Atherton, Chicago, 1972) and Rowell, T., *Social Behavior of Monkeys* (Penguin Books, London, 1972).

[18] Rowell, T., 'Long term changes in a population of Ugandan baboons', *Folia Primat.* **11** (1969), 241–254.

[19] Age specific differences in the ability to memorize have been shown by several workers (e.g., work on *Macaca mulatta* by D. L. Medin and R. T. Davis, 'Memory' pp. 1–47 in *Behavior of Non-Human Primates*, Vol. 5., A. M. Schrier and F. Stollnitz (eds.) (Academic Press, N.Y., 1974)); but the area is far from clear at the moment.

[20] Wilson, E. O., *Sociobiology: The New Synthesis* (Harvard University Press, Cambridge, Mass., 1975).

[21] In her novel *Fear of Flying* (Holt, Rinehart and Winston, New York, 1973) Erica Jong describes the heroine's mother as follows:

"Re: her interest in predation: she started out, I think, with the normal Provincetown – Art Students League communism of her day, but gradually, as affluence and arteriosclerosis overtook her (together, as is often the case), she converted to her own brand of religion composed of two parts Robert Ardrey and one part Konrad Lorenz. I don't think Ardrey or Lorenz intended what she extracted in their names: a sort of Neo-Hobbesianism in which it is proven that life is nasty, mean, brutish, and short; the desire for status, money and power is universal; territoriality is instinctual; and selfishness, therefore, is the cardinal law of life. (Don't twist what I'm saying, Isadora; even what people call altruism is selfishness by another name!)"

22 See for example: C. Lévi-Strauss, *The Savage Mind* (University of Chicago Press, Chicago, 1966); and Mary Douglas, *Natural Symbols: Explorations in Cosmology* (Pantheon Books, New York, 1970).

23 Chomsky, N., *Language and Mind* (Harcourt, Brace, Jovanovich, New York, 1972).

24 Wittgenstein, L., *Tractatus Logico-Philosophicus*, trans. C. K. Ogden, with introduction by Bertrand Russell (Routledge and Kegan Paul, London, 1958).

25 The interpretation of the final statement in the *Tractatus* in this sense (as opposed to the interpretation which would consider the *Tractatus* to enjoin silence on all but scientists) is convincingly defended by: A. Janik and S. Toulmin, *Wittgenstein's Vienna* (Simon and Schuster, New York, 1973).

26 This has been discussed at greater length in Slobodkin, L. B., 'Is history a consequence of evolution?' *Advances in Ethology* 3 (1978), 233–255.

27 This paper was written in 1974 and revised and accepted for publication in 1975. I have deliberately not attempted to update it, since, on rereading I find that the questions, speculations and particularly the cautionary notes and reservations have not been rendered obsolete by more recent publications.

Specifically Wilson (E. O. Wilson, *On Human Nature*, Harvard University Press, 1978), Dawkins (R. Dawkins, *The Selfish Gene*, Oxford University Press, 1976), Alexander (R. Alexander, *Darwinism in Human Affairs*, University of Washington Press, 1980) and others have added to the framework of nativism the notion that genes, as such, cause the bodies in which they are lodged to tend to act in such a way as to maximize the self-interest of the genes themselves, and even to cause the container organisms to recognize, and behave kindly towards, other organisms that carry genes identical with their own genetic contents. These speculations have been metaphorically suggested by evolutionary analyses – particularly those of population genetics – in much the same way that Freud's early speculations about the importance of the Primal Scene in the development of human behavior were suggested by phylogenetic aspects of evolutionary theory.

The use of either the Freudian or sociobiological speculations as a starting point, permits a kind of explanation of almost any personal or historical scenario. It would not be difficult to analyze the biblical account of the Exodus, and the role of Moses, in sociobiological terms, just as Freud, in *Moses and Monotheism*, provided a psychoanalytic interpretation.

We might also consider, for example, a sociobiological reading of Shakespeare in which Claudius' unsympathetic attitude towards Hamlet may be seen as blatantly equivalent to such phenomena as the killing of pre-existing young by the newly arrived male rats and lions that mate with their mothers. Queen Gertrude's ambivalence towards Hamlet may be related to the abortion of female rats that have found new mates. Laertes' fury at the death of his sister, although he hardly seems that upset at the death of his father, is easily explained by the fact that his sister had a much higher reproductive potential at the time of her death than did his father.

As an exercise I leave the question of explaining Hamlet's obsession with his mother's sex life in terms of Trivers ('Parent-offspring conflict', *American Zoologist* **14** (1974), 249–264) rather than Freud.

A central contention of my paper was that human capacity for fantasy, and human consciousness itself, while obviously results of evolution, are not to be simplistically explained. Nevertheless, regardless of whether or not they are conveniently explainable, human fantasy and speculation are major forces in human history, both that of individuals and that of nations. Their effect on history is to undermine, rather than to bolster, deterministic approaches to historiography. (Cf. L. B. Slobodkin, 'Is history a consequence of evolution?' *Perspectives in Ethology* **3** (1978), 233–255.)

Freudian psychoanalytic theory and Wilsonian-Hamiltonian-Dawkinsesque sociobiology share a claimed capacity to generate explanations of all history. In fact this capacity would not have been diminished had history been quite different, and in that sense they provide no explanation at all.

I therefore do not have a sense of having all my questions, since 1975, answered for me by subsequent theoretical breakthroughs. I feel the questions of the paper still stand as questions. The suggested scenario for the development of consciousness remains a plausible one, by the classical criteria of evolutionary plausibility.

LANGDON WINNER

TECHNOLOGIES AS FORMS OF LIFE

Perhaps the most accurate observation one can make about the philosophy of technology is that there really isn't one. At least if we look at the writings of the two sorts of people who might be expected to have been interested in the topic – philosophers and engineers – we find little attention to questions about the character and meaning of technology in human life. For example, the six-volume *Encyclopedia of Philosophy*, a recent attempt at a compendium of major questions in the traditions of philosophical discourse, contains no entry whatsoever under the category 'technology.'[1] Neither does that work contain enough material under possible alternative headings to enable anyone to piece together an idea of what a philosophy of technology might look like.

This is not to say that there is no literature by philosophers in this area. Mitcham and Mackey's *Bibliography of the Philosophy of Technology* lists well over a thousand books and articles in several languages by nineteenth and twentieth century authors.[2] But to begin reading through the material listed shows, in my view, little of enduring intellectual substance. The best writing on this theme comes to us from a few powerful thinkers who have encountered the subject in the midst of much broader and ambitious investigations – for example, Marx in the development of his theory of historical materialism or Heidegger as an aspect of his ontology. It may well be that the philosophy of technology is best seen as derivative from more fundamental questions. For despite the fact that nobody would deny its importance to an adequate understanding of the human condition, technology has never joined epistemology, metaphysics, esthetics, ethics, law, science, and politics as a fully respectable topic for philosophical inquiry.

Engineers have shown little interest in filling this void. Except in the airy pronouncements of the presidential addresses of various engineering societies affirming the contributions of a particular technical vocation to established social institutions, engineers generally appear unaware of any philosophical questions that their work might entail. As a way of starting (or perhaps, more accurately, stopping) a conversation with my colleagues and students in engineering, I sometimes ask: "What are the founding principles of your discipline?" The question is always greeted with puzzlement. Even as I

R. S. Cohen and M. W. Wartofsky (eds.), Epistemology, Methodology and the Social Sciences, 249–263.

explain what I am after, namely, an intelligible account of the nature and significance of the branch of engineering in which they are involved, the question still means nothing to them. While there are a scant few individuals who do raise important first questions about their technical professions, they are usually seen by their colleagues as dangerous cranks and radicals. If Socrates' suggestion that "the unexamined life is not worth living" still holds, this is news to most engineers.

As a starting point, one can say that a philosophy of technology ought to examine critically the nature and meaning of artificial aids to human activity. In point of fact, however, inquiries of that sort are largely missing from modern thought. Our culture is by now firmly based in the success of a wide variety of technical aids – the new kinds of apparatus, techniques and systems that have been devised during the past two centuries. Yet this technical culture has remained remarkably unself-reflective about its own foundations. There are many reasons that might be cited to account for this state of affairs. But certainly among the most important is simply the fact that in our traditional view of the matter, the human situation with regard to technology seems just too obvious to merit serious reflection.

The deceptively reasonable notion that we have inherited from much earlier and less complicated times divides the range of possible concerns about technology into two basic categories: *making* and *use*. In the first of these, our attention is drawn to the matter of "how things work" and of "making things work." We tend to think that this is a sphere of fascination for certain people in certain situations, but not for others. Thus, "how things work" is the domain of inventors, technicians, engineers, repairmen, and the like who prepare artificial aids to human activity and keep them in good working order. Those not directly involved in the various spheres of "making" are thought to have little interest in or need to know about the materials, principles, or procedures found in those spheres.

What the others do care about, however, are tools and uses. This is understood to be a straightforward matter. Once things have been made, we interact with them on occasion to achieve purposes we intend. One picks up a tool, uses it, and puts it down. One picks up a telephone, talks on it, and then does not use it for a time. A person gets on an airplane, flies from point A to point B, and then gets off. The proper interpretation of the meaning of technology in the mode of use seems to be nothing more complicated than an occasional, limited and non-problematic interaction. The grammar of the notion of 'use' includes aspects that enable us to interpret what technologies mean within the realm of moral discourse. Tools are the kinds of things that can be "used well or poorly" and for "good or bad purposes;" I can use my knife to slice a loaf of bread or to stab the next person that walks by. Because technological objects and processes have a promiscuous utility, they are taken to be fundamentally neutral as regards their moral standing.

The conventional idea of what technology is and what it means, an idea powerfully reinforced by certain terms and grammars that we employ in ordinary speech, has to be overcome if a critical philosophy of technology is ever to get under way. A crucial weakness of the conventional notion is that it disregards the many ways in which technologies provide structure for human activity. Since, according to accepted wisdom, patterns that take shape in the sphere of 'making' are of interest to practitioners alone, and since the very essence of 'use' is its occasional, non-structuring occurrence, any attention to form and pattern in technology seems irrelevant.[3]

Yet if the experience of the past two centuries shows us anything, it is certainly that technologies are not only aids to human activity, but also powerful occasions for reshaping that activity. In no area of inquiry is this fact more important than in my own discipline, the study of politics. Since the mid-nineteenth century, highly sophisticated systems of artifice have begun to compete with political institutions as sources of power, order, authority, and loyalty in ways previously unknown. In this new situation many of the most important questions in the tradition of Western political thought simply cannot be addressed without considerable attention to the structures of technology – the various kinds of techniques, apparatus, and systems – that have been added to our world. As a society builds factories, electrical power networks, transportation systems, and the like it is also rebuilding its framework of political order and of citizenship.[4] Thus, the kinds of things that we are apt to see as mere technological entities become much more interesting and problematic if we can begin to notice the ways in which they are broadly involved in the conditions of moral and political life.

It is true, of course, that recurring patterns of life's activities (whatever their origin) tend to become unconscious for us, taken for granted. Thus, we do not pause to reflect upon how we speak a language as we are doing so, or the steps we go through in taking a shower when we are actually in the stall. One point at which we may become aware of a pattern of activity taking shape in the social space of a technical innovation, however, is the first time we encounter that new pattern. An opportunity of that sort happened to me last year at the close of the spring semester. A student in one of my classes came to my office on the day that term papers were due and told me that his essay would be late because "it crashed this morning." I immediately interpreted this as a 'crash' of the conceptual variety, a flimsy array of arguments and observations that eventually collapses under the weight of its own ponderous absurdity. Indeed, many of my own projects have 'crashed' in exactly that manner. But this was not the kind of mishap that had befallen this particular fellow. He went on to explain that his paper had been composed on a computer console and that it had been stored in a time-shared computer memory. In the way such things work, it sometimes happens that the machine 'goes down' or 'crashes,' thus making everything that happens in and around it stop until the computer can be 'brought up,' that is, repaired.

As I listened to the student's explanation, I realized that he was telling me about the facts of a particular form of activity in modern life in which he and others similarly situated were already involved and that I had better get used to. I remembered J. L. Austin's little essay, 'A Plea for Excuses' and noticed that the student and I were negotiating one of the boundaries of contemporary moral life – when and how one gives and accepts an excuse in a particular situation.[5] He was, in effect, asking me to recognize a new world of parts and pieces and of practices and expectations that hold in that world. From then on, a knowledge of the situation would be included in my understanding of not only 'how things work' in the realm of computers, but also of how we do things as a consequence, including which rules are reasonable to follow when the machines break down.

Another illustration can be seen in the experience of two men traveling down in the same direction along a street on a peaceful sunny day, one of them afoot and the other driving an automobile. In this situation, the pedestrian has a certain flexibility of movement; he can pause to look in a shop window, speak to passersby and reach out to pick a flower from a sidewalk garden. The driver, although he has the potential to move much faster, is constrained by the closed space of the automobile, the physical dimensions of the highway and by the rules of the road. His realm is spatially structured by his intended destination, by a periphery of more-or-less irrelevant objects (that can become scenery for occasional side glances) and by more important objects of various kinds – moving and parked cars, bicycles, pedestrians, street signs, etc., that stand in his way. Since the first rule of good driving is 'Avoid hitting things,' the immediate world of the motorist becomes a field of obstacles.

Imagine a situation in which the two persons are next-door neighbors. The man in the automobile observes his friend strolling along the street and wishes to say hello. He slows down, honks his horn, rolls down the window, sticks out his head, and shouts across the street. More likely than not, the pedestrian will be startled or annoyed by the sound of the horn. He looks around to see what's the matter and tries to recognize who can be yelling at him across the way. "Can you come to dinner Saturday night?" the driver calls out over the street noise. "What?" the pedestrian replies, straining to understand. At that moment another car to the rear begins honking to break up the temporary traffic jam. Unable to say anything more, the driver moves on.

What we see here is an automobile collision of sorts, although not one that causes bodily injury. It is a collision between the *world* of the driver and that of the pedestrian. The attempt to extend a greeting and invitation, ordinarily a simple gesture, is complicated by the presence of a technological device and its standard operating conditions. The possibility of communication between the two men is shaped by the inherent incompatibility of the form of life known as walking and a much newer one called automobile driving. Knowing how automobiles are made, how they work, and how they are used, knowing about traffic laws and urban transportation policies, does little to help us understand how automobiles shape experiences like these and what meaning they have for the textures of modern life. It would be valuable to have a phenomenology of technical practice to help us account for the ways, both obvious and subtle, that everyday life is altered by contact with the new procedures, instruments and systems of modern technics; but as yet we have none.

Any number of examples from contemporary experience could be drawn upon to demonstrate how technological development involves the restructuring of the forms of everyday activity – of individual habits, of perceptions, of concepts of self, of social relationships, and of the moral boundaries that govern our understanding of proper conduct. Technologies of various kinds – apparatus, techniques, and man-machine systems – can be seen as templates upon which rennovated forms of modern life take shape, not in any strictly deterministic manner but in ways that allow for varying degrees of improvisation. The concept of determinism here is much too strong and sweeping in its implications. Being saddled with it is much like having only the concept of rape to describe all instances of sexual intercourse. Yet the problem that the idea of technological determinism raises is still an important one.

As a society engages in technological innovation, it does in effect settle many questions without appearing to do so. Vast alterations in the structure of our common world are undertaken with little attention to what those alterations mean. Judgments in this area are, typically, made on extremely narrow grounds. We pay attention to such matters as whether or not the thing in question will serve a particular need, perform efficiently, make a profit, or provide a convenience. Eventually, the broader significance of the structures created may become clear, often as a series of so-called 'unintended consequences.' Such was the case in miniature when the student explained how his paper had crashed or, on a larger scale, in 1973 when the OPEC nations explained what had happened to the petroleum. But it seems characteristic of the human involvement with technology that we are seldom inclined to examine, to discuss, or to judge proposed innovations with a broad awareness of what our decisions entail. In this realm, we repeatedly enter into a series of social contracts, as it were, the terms of which are revealed after the signing. Thus, the issue is not so much that of technological determinism but, rather, of what might be called technological somnambulism – how we so willingly sleepwalk through the process of reconstituting the conditions of human existence.

Recently, there has arisen in social science something called 'technology assessment' or 'social impact analysis' that tries to respond to the element of recurring surprise in technological innovation. The main shortcoming of this kind of work is that it sees technological change as a cause and everything that follows from it as an effect or, to employ a term now in vogue, an 'impact.' A more adequate view, I believe, would notice that as technologies are 'made' and 'used,' significant alterations in patterns of human activity are already taking place. Thus, the construction of any technical system that involves human beings as operating parts amounts to a partial reconstruction of social roles and relationships. Similarly, the very act of using the kinds of machines, techniques, and systems available to us generates patterns of activities and expectations that soon become 'second nature' to us. We do indeed 'use' telephones, automobiles, and electric lights in the conventional sense of picking them up and putting them down. But our world soon becomes one in which telephony, automobility, and electric lighting are forms of life in the sense that life would scarcely be thinkable without them.

My choice of the term 'forms of life' in this context borrows from Wittgenstein's use of that concept in *The Philosophical Investigations*. In his later writing, Wittgenstein sought to overcome an extremely narrow view of the structure of language then popular among philosophers, a view that held language to be primarily a matter of *naming* things and events. Pointing to the richness and multiplicity of the kinds of expression or 'language games' that are a part of everyday speech, Wittgenstein argued that "the speaking of language is part of an activity, or of a form of life."[6] He gave a variety of examples – the giving of orders, speculating about events, guessing riddles, making up stories, forming and testing hypotheses, and so forth – to indicate the wide range of language games involved in various 'forms of life.' Whether he meant to suggest that these are patterns that occur naturally to all human beings or that they are primarily cultural artifacts that can change with the time and setting is a question open to dispute. Hanna Pitkin argues that "Wittgenstein's presentation seems to blur the line between natural and cultural conventions."[7] For my purposes here, however, what matters is not the ultimate philosophical status of Wittgenstein's concept but its suggestiveness in helping us to overcome another widespread and extremely narrow conception, our normal understanding of the place of technologies in human activity.

As they become woven into the texture of everyday life, the devices, techniques and systems that we adopt become a part of our very humanity. In an important sense we become the beings who work on assembly lines, who talk on telephones, who do our figuring on pocket calculators, who eat processed foods, and who clean our homes with powerful chemicals. Of course, working, conversing, figuring, eating, cleaning, and such things have all been parts of human existence for a very long time. But technological innovations can radically alter these common patterns and on occasion generate whole new ones, often with surprising results. For example, none of those who worked to perfect the technology of television in its early years and few of those who brought television sets into their homes ever intended the device to be employed as the universal babysitter. That, however, has become one of TV's most common fucntions in contemporary society.

Most such changes in patterns of everyday life can, of course, be recognized as versions of patterns known before. Parents have always had to entertain and instruct children and to find ways of keeping the little ones out of their hair. Having youngsters watch four hours of color television cartoons on Saturday is, in one way of looking at the matter, merely a new method for handling this task, although the 'merely' is of no small significance. It is important to ask where, if at all, modern technologies have added *fundamentally new* activities to the range of things human beings do and, even beyond that, where and how the innovations of science and technology have begun to alter the very *conditions of life* itself. Is computer programming only a powerful recombination and reshaping of activities that human beings have known for ages – doing mathematics, listing, sorting, planning, organizing, etc. – or is it something that in a true sense is unprecedented? Are the ways of life represented by industrialized agri-business simply revised versions of ways of life traditionally followed since the development of agriculture or do they amount to a new phenomenon?

Certainly, there are some accomplishments of modern technics, manned air flight, for example, that are clearly altogether novel. Flying in airplanes is not just another version of modes of travel previously known; it is something new. What enables us to distinguish mere modifications of our activity from fundamentally new forms of life is, it seems to me, a judgment about whether or not a new device, technique, or system provides something unprecedented to the range of human activities and experiences. Although manned flight as envisioned by the imagination is as old as the story of Daedalus and Icarus, it took a certain kind of modern machinery to realize the dream in actual experience.

Beyond those *activities* that can truly be called novel, however, lie certain kinds of changes now on the horizon that would amount to a fundamental alteration of the *conditions* of human life itself.[8] If it becomes possible to significantly alter the physical structure of human beings through genetic manipulation, or if the possibility of founding permanent settlements situated in outer space were to be realized, then the question of what it means to be human and what constitutes 'the human condition' would become critical questions whose answers are far from obvious. All bets would be off. Speculation about such matters is now largely the work of science fiction whose notorious perversity as a literary genre indicates the troubles that lie in wait when we begin thinking about becoming creatures fundamentally different from any that the earth has seen or of living under conditions fundamentally different from those that life on earth has made available.

But on the whole, it is evident that most of the alterations that occur in the wake of technological innovation are actually variations within very old patterns. Wittgenstein's philosophically conservative maxim – "What has to be accepted, the given, is – so one could say – *forms of life*."[9] – could well be the guiding maxim of a phenomenology of technical practice. Thus, asking a question and awaiting an answer, a form of life we all know well, is much the same activity whether it is a person one is confronting or a computer console. This is not to say that there are no significant differences between a computer console and a person. But the forms of life we mastered before the coming of this new instrument continue to shape our expectations of the thing even though it is distinctly non-human. Recognition of this state of affairs appears in the often pathetic attempts to 'humanize' computers by having them print out 'Hello,' when the user logs in, or having them respond with witty remarks when the user makes an error. We carry with us certain structured anticipations about entities that appear to participate, if only minimally, in forms of life and associated language games that are parts of human culture. This fact is an important source of confusion among those who now predict great advances in 'artificial intelligence' on the basis of certain impressive examples of computer performance. A showroom dummy may wink at me, but that is no reason to fall in love with the thing.

The interpretation of the meaning of technology I am offering here is also similar to Marx's conception of the relationship of human individuality and material conditions of production introduced in Part I of *The German Ideology*. "The way in which men produce their means of subsistence," he explains, "depends first of all on the nature of the means of subsistence they actually find in existence and have to reproduce. This mode of production must not be considered simply as being the reproduction of the physical existence of the individuals. Rather it is a definite form of activity of these individuals, a definite form of expressing their life, a definite *mode of life* on their part. As individuals express their life, so they are."[10] The concept of production that Marx employs here is a very broad and suggestive one. While he clearly points in the first instance to means of production that sustain life in an immediate sense, his view extends to a general understanding of what it means to be a human individual in a world of diverse natural resources, tools, machines, products for consumption, and productive social relationships. The notion is clearly not one of occasional human interaction with devices and material conditions that leave individuals untouched. By changing the world of material things, Marx observes, we also change ourselves. In this process human beings do not stand at the mercy of a great deterministic punch press that cranks out precisely shaped persons at a certain rate during a given historical period. Instead, the situation Marx describes is one in which individuals are actively involved in the daily creation and re-creation, production and reproduction, of the world in which they live. Thus, as they employ tools and techniques, work in social arrangements of production, make and consume kinds of products, and as they adapt their behavior to the material conditions they encounter in their natural and artificial environment, individuals realize possibilities for human existence that are inaccessible in more primitive modes of production.

Marx expands upon this idea in 'The Chapter on Capital' in the *Grundrisse*. The development of forces of production in history, he argues, holds the promise of the development of a many-sided individuality in all human beings. Capital's unlimited pursuit of wealth leads it to develop the productive powers of labor to a stage "where the possession and preservation of general wealth require a lesser labour time of society as a whole, and where the labouring society relates scientifically to the process of its progressive reproduction, its reproduction in constantly greater abundance . . . "[11] This movement toward a general form of wealth "creates the material elements for the development of the rich individuality which is all-sided in its production as in its consumption, and whose labour also therefore appears no longer as labour, but as the full development of activity itself . . . "[12] As one has access to the tools and materials of woodworking, a person can develop himself through the activities of carpentry; as one has access to instruments and techniques for making music, one can become (in that aspect of one's life) a musician. Marx's ideal here, a kind of materialist humanism, anticipates that in a properly structured society under modern conditions of production, people would engage in a very wide range of activities that enrich their individuality along many dimensions.[13]

As applied to our understanding of technology, the philosophies of Wittgenstein and Marx are similar in the sense that they draw our attention to the fabric of everyday actions and interactions in which human beings participate and from which they derive meaning. Wittgenstein's emphasis upon forms of life points to the vast multiplicity of cultural patterns that comprise our common world. The helpful tendency of this mode of analysis is to pay close attention to parts and pieces, the minutiae of daily existence. For example, by noticing 'what we say when' and what we actually do in the matter of telling time, it should be possible to appreciate how, if at all, the adoption of digital clocks and watches has altered our sense of time. Marx, on the other hand, directs our attention to much larger historical patterns, stages in the development of agriculture and industry, for example, in which a particular mode of production is seen as the form in which individuals express their life. Marx seeks to explain relationships within these large patterns, e.g., relationships between different social classes in a given stage of the development of productive forces. But as the chapter on 'Machinery and Large-Scale Industry' in *Capital* demonstrates, his mode of interpretation includes a place for a more microscopic treatment of the role of specific technologies in human experience.[14]

My invocation of Wittgenstein and Marx in this context, however, is not a suggestion that either one or both in combination give us an adequate standpoint for raising the critical questions for a philosophy of technology. In proposing an attitude in which forms of life must be accepted as "the given," Wittgenstein decides that philosophy "leaves everything as it is."[15] Although Wittgensteinians are eager to point out that this does not necessarily predispose the philosopher to conservatism in a social or political sense, it does seem that as applied to the study of forms of life in technological innovation, Wittgenstein leaves us with little more than a passive traditionalism. If one hopes that one's investigations here would be useful not only in the interpretation of technological phenomena but also serve as the grounds for positive judgments and action, then Wittgensteinian philosophy as such leaves much to be desired.

Marx and, for that matter, Marxism in virtually all of its forms present shortcomings of another kind. This mode of understanding points us toward the identification of all but inevitable historical tendencies that promise human emancipation at some point. As the forces of production and social relations of production develop and as the proletariat makes its way toward revolution, Marx and his orthodox followers are willing to allow capitalist technology, e.g., the factory system, to develop to their furthest extent. Marx and Engels scoffed at the utopians, anarchists and Romantic critics of industrialism who thought it possible to make moral and political judgments about the course a technological society ought to take and to influence that path through the application of philosophical principles. In its own way, then, the Marxist theory of *praxis* contains a passivity almost as debilitating as the Wittgensteinian decision to leave "everything as it is." The famous eleventh thesis on Feuerbach – "The philosophers have only interpreted the world in various ways; the point, however, is to *change* it."[16] – conceals an important qualification: that judgment, action and change are ultimately products of history. As regards technological development, Marxism anticipates a history of increasing productivity in which attempts to propose limits have no place.

My use of Marx and Wittgenstein, then, is simply to indicate a way in which the now dominant and debilitating conceptions about 'making' and 'use' on the one hand and technological 'impacts' on the other might be overcome. From the starting point I have suggested, there are a number of directions in which the inquiry could be carried. One might, for instance, take up the study of human responses to technical objects, processes and systems employing the methods of contemporary anthropology or 'ethnomethodology' as a means of interpretation. One could examine in detail the ways in which our everyday activities, sensibilities, and understandings change when confronted with new technological templates. In this regard Joseph Weizenbaum's fascinating account of the 'compulsive programmer,' the person who sleeps on a cot next to the computer and sends out for sandwiches as he endlessly debugs his program, provides a model of how novel activities and passions generated in the course of technological change might be described.[17] Richard Rabinowitz's *Soul, Character and Personality*, an historical reconstruction of the transformation of human experience during the industrial revolution in America, presents a richly detailed approach to the past in this focus of concerns.[18] Rabinowitz is able to demonstrate a range of alterations in the ways in which evangelical Christians thought about their God, their world, and their own identities, alterations that were in important ways a response to a changing material culture – the newly emergent industrial, commercial, technological world taking shape around them. For both Weizenbaum and Rabinowitz the phenomena observed are not mere 'effects' or 'impacts' that follow variations in the technological sub-structure. What we see are human individuals actively renovating what they are doing (and how they understand what they are doing) in a world in which certain possibilities have opened up and others have suddenly closed.

My own interest in the subject carries me along a somewhat different path toward a different set of questions. The forms of life that interest me are ultimately political ones, the forms of participation and citizenship that exist in a given community at a given time. Insofar as technologies are templates which influence the shape and texture of political life, that is a topic of concern for the theory of politics. The traditional domain of political theory includes not only the study of how things are, have been and are becoming, but also the inquiry into how things ought to be. Studies that rest content with the business (and it is increasingly a business) of describing, interpreting, and explaining technological change leave their most important task unfinished. As the political imagination confronts technologies as forms of life, it should be able to say something about the choices (implicit or explicit) made in the course of technological innovation and the grounds for making those choices wisely. This must be done in the face of a general practice in modern society which, under the universal influence of the doctrine of Progress, has taken such questions lightly or ignored them altogether. Rather than passively monitor the spectacle of transformations past and present, political theory ought to articulate the meaning of technological change for our common life and offer guidance in the matter of choosing among alternatives.

In this light an important issue that presents itself has to do with how one might understand the political significance of technological *design*. Is it possible to say with any precision that technological apparatus and technological systems embody political ideas or embody the distinctive forms of political life in a society? Can we say of a given entity – a transportation or communications system, for example – that its design has certain properties, political properties, that another design in roughly the same domain of practice does not have? The choice of a technological template properly involves considerations far beyond the threadbare criteria of judgment – utility, efficiency, economy, convenience, competitiveness – that the modern age has obsessively applied. To find ways of talking about a richer, more meaningful set of criteria and how they might influence not only 'yes or no' choices but also choices among possible alternative designs is a crucial issue for the philosophy of technology (considered as the political philosophy of technology) to address.

It is true that not every technological innovation embodies choices of great significance. Some developments are more or less innocuous; many are socially and politically trivial. But in general, where one finds substantial changes being made in what people are doing and at a substantial investment of social resources, then one can usually find technologically embodied answers to questions about power, authority, order, freedom and the other central themes in Western political thought.

In this context the ironies of technological somnambulism are especially poignant. If, for example, a society has never made an explicit decision about the centralization of political control, in fact even if that society has declared its opposition to such centralization, it may nevertheless happen that the adoption of a centralized design in its systems of communication or energy production in effect settles that issue. If a society has never made the explicit choice to concentrate power in the hands of a relative few and even if that society supposes that its political institutions contain guarantees of dispersed power, it may nonetheless happen that socio-technical designs that concentrate power over major systems of production or service can have the effect of concentrating power in a more general sense. To put it differently, while people may be unwilling to accept the institution of a great tyranny, they may perhaps embrace that same institution if it is offered as a wonderful home convenience or as the next step in a course of technical necessity.

It goes without saying that not everyone is equally somnambulistic with regard to the political significance of technological design. Some historical actors – the planners of the Bay Area Rapid Transit system in San Francisco, for instance – have been very clear about what was at stake in things others saw as innocuous blueprints and diagrams.[19] There are good reasons why the largest banks in the United States are enthusiastic about a particular model of electronic funds transfer. Most of us, however, experience such things as BART and EFT as yet more in a series of amazing developments that have rendered certain things we used to do 'obsolete.' A deceptive quality of technical objects and processes – their promiscuous utility, the fact that they can be 'used' in this way or in that – blinds us to the ways in which they structure what we are able to do and the ways in which they settle important issues *de facto* without appearing to do so. Thus, for example, the freedom we enjoy in the realm of 'use' is mirrored in our extreme dependency upon vast, centralized, complicated, remote, and increasingly vulnerable artificial systems.

It may seem ironic, then, that what I propose is actually a return to the sphere of *making* with a new set of concerns in view. If the sterility of modern thinking with regard to technology is to be overcome, the subject must be re-opened in the light of knowledge-constitutive interests, to employ the familiar terms of Jürgen Habermas, that are at once 'practical' and 'emancipatory'.[20] From the perspective offered by Hannah Arendt's philosophy of the human condition, the most interesting aspect of the *vita activa* at present is that of 'work' or 'fabrication', the activity through which *homo faber* strives to erect a durable home on earth.[21] Arendt understood that activity of this kind involves the combined work of artists, craftsmen, historians, poets, architects, engineers, and constitution makers. Arendt herself, however, did not go very far in applying this insight to the 'work' of modern material culture. I am suggesting that it is now useful to think about technological design features in roughly the same way that the legislators of the ancient world or the eighteenth-century philosophers pondered the structural characteristics of political constitutions. Technologies provide frameworks of order for the modern world. As such, it now makes sense to try to understand the forms of authority, justice, public good, and freedom that their order entails.

University of California, Santa Cruz

NOTES

[1] *The Encyclopedia of Philosophy*, Paul Edwards, editor-in-chief, 8 vols. (Macmillan Publishers, New York, 1967).

[2] *Bibliography of the Philosophy of Technology*, Carl Mitcham and Robert Mackey, eds. (University of Chicago Press, Chicago, 1973).

[3] An excellent corrective to the general thoughtlessness about 'making' and 'use' is to be found in Carl Mitcham's 'Types of Technology' in *Research in Philosophy and Technology*, Paul Durbin, ed. (JAI Press, Inc., Greenwich, Connecticut, 1978), pp. 229–294.

[4] For examples of this phenomenon described historically, see Alfred D. Chandler's *The Visible Hand: The Managerial Revolution in American Business* (Harvard University Press, Cambridge, Mass., 1977).
[5] J. L. Austin, *Philosophical Papers* (Oxford University Press, Oxford, 1961), pp. 123–152.
[6] Ludwig Wittgenstein, *Philosophical Investigations*, G. E. M. Anscombe, trans., third edition with English and German indexes (Macmillan, New York, 1958), p. 11e.
[7] Hanna Pitkin, *Wittgenstein and Justice: On the Significance of Ludwig Wittgenstein for Social and Political Thought* (University of California Press, Berkeley, Los Angeles and London, 1972), p. 293.
[8] Hannah Arendt, *The Life of the Mind, Vol. II: Willing* (Harcourt Brace Jovanovich, New York and London, 1978).
[9] *Philosophical Investigations*, p. 226e.
[10] Karl Marx and Friedrich Engels, *The German Ideology*, in *Collected Works*, vol. 5 (International Publishers, New York), p. 31.
[11] Karl Marx, *Grundrisse*, translated with a foreword by Martin Nicolaus (Penguin Books, Harmondsworth, England, 1973), p. 325.
[12] *Ibid.*
[13] An interesting discussion of Marx in this respect is Kostas Axelos' *Alienation, Praxis and Technē in the Thought of Karl Marx*, translated by Ronald Bruzina (University of Texas Press, Austin, Texas, 1976).
[14] Karl Marx, *Capital*, Vol. 1, translated by Ben Fowkes with an introduction by Ernest Mandel (Penguin Books, Harmondsworth, England, 1976), Ch. 15.
[15] *Philosophical Investigations*, p. 49e.
[16] In *Collected Works*, Vol. 5, p. 8.
[17] Joseph Weizenbaum, *Computer Power and Human Reason: From Calculation to Judgment* (Freeman, San Francisco, 1976), Ch. 4.
[18] Richard Rabinowitz, 'Soul, Character and Personality: The Transformation of Personal Religious Experience in New England, 1790–1860' (Ph.D. dissertation, Harvard University, 1977, unpublished).
[19] Steven Zwerling, *Mass Transit and the Politics of Technology: A Study of BART and the San Francisco Bay Area* (Praeger, New York, 1974).
[20] Jürgen Habermas, *Knowledge and Human Interests*, translated by Jeremy J. Shapiro (Beacon Press, Boston, 1971).
[21] Hannah Arendt, *The Human Condition* (University of Chicago Press, Chicago, 1958), Ch. IV.

INDEX OF NAMES